T0325685

Solid–Solid, Fluid–Solid, Fluid–Fluid Mixers

There are no such things as applied sciences,
only applications of science.
Louis Pasteur (11 September 1871)

Dedicated to my wife, Anne, without whose unwavering support, none of this
would have been possible.

Industrial Equipment for Chemical Engineering Set

coordinated by
Jean-Paul Duroudier

Solid–Solid, Fluid–Solid, Fluid–Fluid Mixers

Jean-Paul Duroudier

First published 2016 in Great Britain and the United States by ISTE Press Ltd and Elsevier Ltd

ISTE Press Ltd
27-37 St George's Road
London SW19 4EU
UK

Elsevier Ltd
The Boulevard, Langford Lane
Kidlington, Oxford, OX5 1GB
UK

www.iste.co.uk

www.elsevier.com

For information on all our publications visit our website at http://store.elsevier.com/

British Library Cataloguing-in-Publication Data
A CIP record for this book is available from the British Library
Library of Congress Cataloging in Publication Data
A catalog record for this book is available from the Library of Congress
ISBN 978-1-78548-180-2

Printed and bound in the UK and US

Contents

Preface

The observation is often made that, in creating a chemical installation, the time spent on the recipient where the reaction takes place (the reactor) accounts for no more than 5% of the total time spent on the project. This series of books deals with the remaining 95% (with the exception of oil-fired furnaces).

It is conceivable that humans will never understand all the truths of the world. What is certain, though, is that we can and indeed must understand what we and other humans have done and created, and, in particular, the tools we have designed.

Even two thousand years ago, the saying existed: "faber fit fabricando", which, loosely translated, means: "*c'est en forgeant que l'on devient forgeron*" (a popular French adage: *one becomes a smith by smithing*), or, still more freely translated into English, "practice makes perfect". The "artisan" (faber) of the 21st Century is really the engineer who devises or describes models of thought. It is precisely that which this series of books investigates, the author having long combined industrial practice and reflection about world research.

Scientific and technical research in the 20th century was characterized by a veritable explosion of results. Undeniably, some of the techniques discussed herein date back a very long way (for instance, the mixture of water and ethanol has been being distilled for over a millennium). Today, though, computers are needed to simulate the operation of the atmospheric distillation column of an oil refinery. The laws used may be simple statistical correlations but, sometimes, simple reasoning is enough to account for a phenomenon.

Since our very beginnings on this planet, humans have had to deal with the four primordial "elements" as they were known in the ancient world: earth, water, air and fire (and a fifth: aether). Today, we speak of gases, liquids, minerals and vegetables, and finally energy.

The unit operation expressing the behavior of matter are described in thirteen volumes.

It would be pointless, as popular wisdom has it, to try to "reinvent the wheel" – i.e. go through prior results. Indeed, we well know that all human reflection is based on memory, and it has been said for centuries that every generation is standing on the shoulders of the previous one.

Therefore, exploiting numerous references taken from all over the world, this series of books describes the operation, the advantages, the drawbacks and, especially, the choices needing to be made for the various pieces of equipment used in tens of elementary operations in industry. It presents simple calculations but also sophisticated logics which will help businesses avoid lengthy and costly testing and trial-and-error.

Herein, readers will find the methods needed for the understanding the machinery, even if, sometimes, we must not shy away from complicated calculations. Fortunately, engineers are trained in computer science, and highly-accurate machines are available on the market, which enables the operator or designer to, themselves, build the programs they need. Indeed, we have to be careful in using commercial programs with obscure internal logic which are not necessarily well suited to the problem at hand.

The copies of all the publications used in this book were provided by the *Institut National d'Information Scientifique et Technique* at Vandœuvre-lès-Nancy.

The books published in France can be consulted at the *Bibliothèque Nationale de France*; those from elsewhere are available at the British Library in London.

In the in-chapter bibliographies, the name of the author is specified so as to give each researcher his/her due. By consulting these works, readers may gain more in-depth knowledge about each subject if he/she so desires. In a reflection of today's multilingual world, the references to which this series points are in German, French and English.

The problems of optimization of costs have not been touched upon. However, when armed with a good knowledge of the devices' operating parameters, there is no problem with using the method of steepest descent so as to minimize the sum of the investment and operating expenditure.

Stirring in a Vat:
Homogenization of Pasty Products

1.1. Principles

1.1.1. *Types of stirrers*

Traditional stirring in a vat uses a stirrer made up of a shaft at the end of which a thruster is attached. Such a configuration is feasible only for liquids whose viscosity is less than 60 Pa.s.

The thruster can cause an axial flow of the liquid. Thus, it could be:

– a marine propeller;

– a saber-blade propeller;

– an axial turbine (with inclined palettes).

If, on the other hand, the flow engendered is radial, we shall be dealing with:

– a radial turbine

Axial thrusters cause a significant degree of circulation of the liquid in the vat, and it is this which makes them advantageous. On the other hand, radial thrusters are chosen because they rotate quickly and, at their periphery, create a high degree of shear.

1.1.2. *Axial thrusters (circulating)*

The marine propeller includes three blades. The axial displacement corresponding to a rotation of 2π radians is known as the "step of the propeller"; the relative step is the ratio of the step to the diameter of the propeller. That ratio is between 0.8 and 1.2. When it is equal to 1, we say that the propeller has a square step. This is the most usual situation.

This thruster is used to cause significant circulation in the vat. The liquid descends along the axis and re-ascends along the walls of the vat. Consequently, this thruster is well suited to homogenize a solution and to suspend a solid or to keep it in suspension. The rotation rate can be chosen between 200 and 1500 rev.mn^{-1}

The viscosity of the product must be no greater than 8 Pa.s when we are using a marine propeller.

The thruster is generally placed at one third of the liquid level from the bottom of the vat, and the ratio d/D is of the order of 1/3.

The saber-bladed propeller or hyper-circulator works on the basis of the same principle as the marine propeller, but its three blades are cut into sheet metal. In comparison to the marine propeller, the hyper-circulator exhibits the following advantages:

– the shear caused is less, because the hyper-circulator turns more freely;

– the weight is less, because the blades are slimmer;

– the manufacturing cost is less;

– the blades can be disassembled;

– for a given degree of circulation, the energy expenditure is less.

A hyper-circulator can be designed with a rotation speed of between 20 and 250 rev.mn^{-1}. The acceptable viscosity for the product may be up to 60 Pa.s, because the blades are slender. It therefore happens that this thruster works in the laminar regime and, this being the case:

$N_p = 32/Re$ (see section 1.2.1)

Another advantage of the hyper-circulator is that, as its diameter is large, the product is set in motion in immediate proximity to the walls of the vat, even if its viscosity is high.

As a hyper-circulator is lighter than a propeller, it is reasonable for its diameter to easily reach up to 70% of the diameter of the vat, so the maximum diameter of the thruster would be 6 m, whereas for marine propellers, it is only 0.8 m. However, the torque that needs to be transmitted by the reducer to the motor is then significant, and that reducer is expensive. However, the additional investment is compensated by the saving in terms of driving energy, because the device turns slowly.

In the conditions of use of a marine propeller, the hyper-circulator only exhibits advantages, but only when the device is large.

The thruster with four palettes inclined at 45° from the horizontal is the less-advanced version of the two discussed above. If d is the diameter of the thruster, the breadth of the palettes is d/5 and their length is d/2. As this thruster has four blades instead of three, it provides better circulation than a hyper-circulator. On the other hand, its energy consumption – i.e. the shearing force engendered – is high.

It is possible to exploit the significant circulation associated with high shear force to disperse an immiscible liquid in another, even if the ratio of the viscosities of the two liquids is up to 10^5 and if the difference between their densities is up to 100 kg.m^{-3}.

1.1.3. Radial thrusters (shearing)

The palette turbine has a disk whose diameter is ¾ that of the full thruster. Affixed to the edge are six vertical rectangular palettes whose length is d/4 and whose height is d/5 (d is the diameter of the full thruster).

The main use of this type of turbine is dispersion of a gas. The gas is injected through a corona shot with holes, laid beneath the turbine, which bursts the rising bubbles. The disk prevents the bubbles from rising along the axis without being divided.

The rotation speed can be chosen between 200 and 1500 rev.mn^{-1}, as is the case with the bar turbines which we are about to examine.

These turbines can only be used in highly-fluid liquids. Indeed, the circulation they cause is only slight, and would be absolutely non-existent in a liquid that is even a little viscous.

Just like the palette turbine, the bar turbine contains a disk whose diameter is equal to 3d/4. At the edge of this disk, there are six welded radial horizontal bars whose length is d/4 and whose cross-section is square, with a side measuring d/20. Three bars are affixed to the top of the disk and they alternate with three bars fixed to the bottom.

This thruster is used to disperse an immiscible liquid into another. On this issue, it must be pointed out that if we wish to disperse a liquid A into a liquid B, both liquids must be arranged in the vat with the interfacial plane clearly identified, and we must immerse the thruster in the liquid chosen to be the continuous phase – here, liquid B. The thruster must be only a short distance – of the order of d/2 – away from the plane of the interface.

It is also possible to use this device to shred and disperse a fibrous solid into a liquid. In this case, we have a grinding turbine.

As the circulating power of shearing thrusters is poor, there is a danger that it will take a long time to obtain a satisfactory liquid–liquid emulsion. For this reason, we have to use an axial turbine.

1.1.4. *Circulation in a stirred vat*

In order to prevent the liquid simply traveling around with the thruster, the vat has four baffles whose breadth is 1/10 the diameter of the vat if the viscosity of the liquid is comparable to that of water. This breadth then decreases as a linear function of the viscosity, and is zero at 10 Pa.s.

For ease of cleaning, the baffles must be a distance equal to half the breadth of the baffles away from the wall.

Figure 1.1 illustrates the circulation of the liquid obtained with an axial thruster.

The points M and N are on a horizontal circle, the fixed disk, in the vicinity of which the liquid's motion is very slight, if not non-existent.

The total circulated flowrate Q_T is the sum of two flowrates:

– the primary flowrate Q_0, exiting directly from the circular area swept by the rotating thruster;

– the flowrate Q_I induced by the above flow:

$$Q_T = Q_0 + Q_I$$

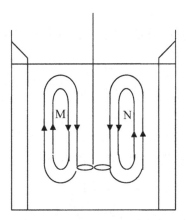

Figure 1.1. *Axial thruster*

Let R_0 represent the radius of the fixed circle and Z_0 the distance from that disk to the bottom of the vat. There are various ways of estimating the total flowrate:

1) By dividing the vat by the plane of the fixed disk:

$$Q_T = 2\pi \int_0^{R_0} V_Z\ RdR = 2\pi \int_{R_0}^{D/2} V_Z\ RdR$$

2) By dividing the vat with a vertical cylinder of radius R_0:

$$Q_T = 2\pi R_0 \int_0^{Z_0} V_R\ dZ = 2\pi R_0 \int_{Z_0}^{Z_T} V_R\ dZ$$

Z_T is the height of liquid in the vat.

The velocities V_R and V_Z are measured by laser Doppler anemometry. This method of investigation also means we can measure not only the mean velocity at a given place, but also the fluctuations in that velocity.

If V_c is the volume of the vat, the time T_c taken to complete a full circulation within the vat is:

$$T_c = V_c / Q_T$$

T_c is the period of circulation.

In the turbulent regime established, the flowrate Q_T is proportional to the speed of rotation of the thruster and to the cube of the thruster's diameter:

$$Q_T = N_Q N d^3$$

N_Q is the reduced flowrate, also known as the "flowrate number".

This reduced flowrate characterizes the type of thruster chosen.

As regards radial thrusters, Figure 1.2 represents the circulation field obtained.

The field obtained is in the shape of a four-leafed clover. In view of the fact that the zones of circulation situated above and below the thruster interact very little at the level of the thruster itself, radial thrusters understandably should be discounted if we want extensive full circulation in the vat.

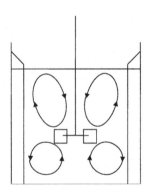

Figure 1.2. *Radial thruster*

Thus, we can say that axial thrusters are circulating thrusters and that radial thrusters are shearing thrusters. Indeed, the shearing effect exerted on the liquid is intense at the edge of radial thrusters.

1.1.5. Maximum shear

Consider (in cylindrical coordinates) the Navier equation pertaining to the orthoradial velocity, as given in Brun *et al.* ([BRU 68] Vol. III p. 219). If we operate in the plane of symmetry of the thruster, if the regime is permanent and as the symmetry is symmetry of revolution, simplifications can be made, and this equation is written:

$$U_r \frac{\partial U_\theta}{\partial r} + \frac{U_r U_\theta}{r} = v\left(\frac{1}{r}\frac{\partial U_\theta}{\partial r} + \frac{\partial^2 U_\theta}{\partial r^2} - \frac{U_\theta}{r^2} \right)$$

This equation takes a completely different form depending on whether we are in a plainly laminar or a plainly turbulent regime.

1) Plainly laminar regime

This is the case if the kinematic viscosity v is greater than 1 m²/s. Only the right-hand side of the previous equation remains, and after integration, it can be written as:

$$\frac{U_\theta}{r} + \frac{\partial U_\theta}{\partial r} = \frac{1}{\tau_0}$$

Let us integrate once more:

$$U_\theta = \frac{r}{2\tau_0} - \frac{K}{r}$$

U_θ increases indefinitely with r, and the product tends to rotate in a clump along with the thruster. However, a kinematic viscosity of 1 m.s^{-2} corresponds to a dynamic viscosity of 100 Pa.s. This type of product is treated in a mixer (which is also useful for products whose viscosity is greater than 200 Pa.s).

2) Plainly turbulent regime

In the turbulent regime, the velocities undergo random fluctuations, and we can posit:

$$U_\theta = \bar{U}_\theta + u_\theta \quad \text{and} \quad U_r = \bar{U}_r + u_r$$

The overlined letters represent mean values over time, and the lowercase letters represent the fluctuations.

As the product is highly fluid ($\upsilon \# 10^{-6}$ m^2/s), only the left-hand side of the Navier equation remains:

$$\left(\bar{U}_r + u_r\right)\frac{\partial\left(\bar{U}_\theta + u_\theta\right)}{\partial r} + \frac{\left(\bar{U}_r + u_r\right)\left(\bar{U}_\theta + u_\theta\right)}{r} = 0$$

Let us take the mean of the terms in this equation over time. The mean of the fluctuations is null, as is their derivative from r. Hence:

$$\bar{U}_r\frac{\partial\bar{U}_\theta}{\partial r} + \frac{\bar{U}_r\bar{U}_\theta}{r} + \frac{\overline{u_r u_\theta}}{r} = 0$$

The maximum shear is then (if r is the half-diameter of the thruster):

$$\frac{\partial\bar{U}_\theta}{\partial r} = -\frac{\bar{U}_\theta}{r} - \frac{\overline{u_r u_\theta}}{\bar{U}_r r}$$

EXAMPLE 1.1.–

By looking at the figure on page 200 of [UHL 66], Vol. I), we can see that, at the periphery of a radial turbine:

$$r = 0.15 \text{ m} \quad \frac{\partial\bar{U}_\theta}{\partial r} = -17 \text{s}^{-1} \quad \frac{\bar{U}_\theta}{r} = 7.8 \text{ s}^{-1} \quad \text{and} \quad \bar{U}_r = 1 \text{ m.s}^{-1}$$

The maximum shear equation gives us:

$$\overline{u_r u_\theta} = 0.15\left(17 - 7.8\right) = 1.38 \text{ m}^2.\text{s}^{-2}$$

Thus, we see that:

– the turbulence is not homogeneous, because the covariance of u_r and u_θ is far from null;

– as the term $u_r u_\theta$ has a value of the same order as the product $\bar{U}_r \bar{U}_\theta$, we can deduce that the fluctuations in velocity are of the same order of magnitude as the mean value of these velocities.

It is customary to define the intensities of turbulence by:

$$I_r = \left(\frac{\overline{u_r^2}}{\bar{U}_r^2} \right)^{1/2} \quad \text{and} \quad I_\theta = \left(\frac{\overline{u_\theta^2}}{\bar{U}_\theta^2} \right)^{1/2}$$

at the bottom of a turbine-stirred vat, the mean intensity of turbulence is already 0.7.

The intensity of turbulence for the covariance is:

$$I_{r\theta} = \left(\frac{\overline{u_r u_\theta}}{\bar{U}_r \bar{U}_\theta} \right)^{1/2} = \left(\frac{1.38}{1 \times 1.13} \right)^{1/2} = 1.1$$

Provided we remain in a genuinely turbulent regime, if we vary the thruster's rotation speed, experience shows that the intensities of turbulence do not vary at all, so we can write:

$$\frac{\partial \bar{U}_\theta}{\partial r} = -\frac{\bar{U}_\theta}{r} \left[1 + \left(I_{r\theta} \right)^2 \right]$$

However, at the immediate edge of the thruster:

$$\bar{U}_\theta = 2\pi r N$$

Thus:

$$\frac{\partial \bar{U}_\theta}{\partial r} = -2\pi N \left[1 + \left(I_{r\theta} \right)^2 \right]$$

The maximum velocity gradient – i.e. the maximum shearing – is proportional to the rotation speed, which, intuitively, was clear all along.

1.1.6. *Discharge pressure of a turbine*

This is the pressure created by the rotation of the turbine at its periphery.

The rate of variation of the kinetic moment of the fluid in relation to the axis of the turbine is equal to the torque of the forces applied to the fluid by the blades of the thruster:

$$C = rQ\rho\omega r$$

Q and ρ are, respectively, the volumetric flowrate and the density of the fluid. The corresponding power is:

$$P = C\omega = Q\rho\omega^2 r^2$$

and the discharge pressure is:

$$\Delta P = \frac{P}{Q} = \rho\omega^2 r^2 = \rho\left(V_\theta\right)^2 = \rho 4\pi^2\left(\frac{Nd}{2}\right)^2 = \rho\pi^2\left(Nd\right)^2$$

We can see that the discharge pressure is proportional to $(Nd)^2$.

EXAMPLE 1.2.–

The turbine cited in [UHL 66] is characterized by:

$$r = 0.15 \text{ m} \quad \text{and} \quad N = 120 \text{ rev.mn}^{-1} = 2 \text{ rev.s}^{-1}$$

Thus:

$$\Delta P = 1000 \times \pi^2 \times \left(2 \times 0.15 \times 2\right)^2$$

$$\Delta P = 3553 \text{ Pa} = 0.035 \text{ bar}$$

This pressure is low, and corresponds to a height of liquid of 35 cm.

NOTE.–

Similarly to with centrifuge pumps, we can define a discharge height:

$$H = \frac{\Delta P}{\rho g} = \frac{\omega^2 r^2}{g}$$

g is the acceleration due to gravity.

1.2. Power consumed and recirculation rate

1.2.1. Dimensionless numbers

1) The Reynolds number

The Reynolds number is the ratio of the stopping pressure (due to the kinetic energy) to the shear stress due to the viscosity.

The stop pressure is the difference between the pressure levels exerted on a membrane depending on whether that membrane is perpendicular or parallel to the flow of the fluid.

$$Re \sim \frac{\dfrac{\rho V^2}{2}}{\mu \dfrac{\partial V}{\partial x}} \sim \frac{\rho N^2 d^2}{\mu \dfrac{Nd}{d}}$$

Indeed, in the turbulent regime, the mean velocities over time at all points in the vat are proportional to the product Nd.

N: rotation frequency: rev.s^{-1}

d: diameter of the thruster: m

Thus:

$$Re = \frac{\rho N d^2}{\mu}$$

2) The flowrate number (reduced flowrate)

This number is defined by:

$$Q_T = N_Q N d^3$$

Q_T: recirculated flowrate: $m^3.s^{-1}$;

N_Q: flowrate number.

In the turbulent regime, the number N_Q is independent of the Reynolds number, and in the transitory regime, it increases along with it.

The flowrate number characterizes the circulating power of the thruster in the vat.

3) The power number

The power imparted to the fluid is the product of the recirculated flowrate Q_T by the pressure engendered by the thruster. We have seen (section 1.1.6) that this pressure is proportional to $(Nd)^2$, and it is also proportional to the density ρ of the liquid.

$$P = \rho Q_T \Delta P \sim \rho N_Q N d^3 (Nd)^2$$

Thus:

$$P = N_P \rho N^3 d^5 \quad (\text{Watt})$$

N_P is the power number. It is also the reduced power.

The power characteristic is a curve or an analytical expression expressing the variations of the power number as a function of the Reynolds number.

Let us cite some orders of magnitude of the power employed (in $kW.m^{-3}$ of liquid):

Homogenization of vats of vegetable oils   0.004

Homogenization of petrol tanks   0.015

Homogenization of milk cans	0.02 to 0.07
Dissolution of washing powder	0.2
Fermentation of polymerization (in emulsion)	0.5 to 2
Incorporation of chocolate into milk	0.8 to 1
Polymerization in suspension	1.2 to 1.3
Dispersion of clay	2

4) Ratio N_Q/N_P

This ratio characterizes the circulation obtained for a given energy expenditure.

5) Mean driving pressure

This is given by (see section 1.1.6)

$$\Delta P = \frac{P}{Q_T}$$

1.2.2. Geometric coefficient C_G

This coefficient multiplies the power number and expresses the influence of three parameters:

– the relative immersion Z_I/d, which is the quotient by the diameter of the thruster of the distance between the center of the thruster and the free surface of the liquid at rest;

– the relative proximity Z_P/d, which is the quotient by the diameter of the thruster of the distance between the center of the thruster and the bottom of the vat;

– the relative separation Z_S/d, which is the quotient by the diameter of the thruster of the distance between the centers of two successive superposed thrusters.

Let us examine the behavior of the geometric coefficient depending on the type of thruster.

1) Hyper-circulator

This thruster turns slowly and the vortex effect is no concern when $Z_I/d >$ 0.25, which is always the case. If Z_p/d is less than 0.5, it is likely that the geometric coefficient C_G will climb slightly above 1.

If we superpose two hyper-circulators, it is again likely that the C_G will be a function of the distance from the thruster to the thruster immediately below it and, by analogy with propellers:

$$C_G = 0.65 \left(\frac{Z_s}{d} \right)^{0.24}$$

2) Marine propeller

For the same reasons as with the hyper-circulator, we can accept that C_G is equal to 1 in all practical conditions of use of a single propeller. The diameter of the propellers is smaller than that of hyper-circulators and, although they rotate more quickly, the vortex effect remains slight, and the immersion has no impact.

For superposed propellers, the C_G found in the work of Bates, reproduced in [UHL 66] can be expressed by:

$$C_G = 0.675 \left(\frac{Z_s}{d} \right)^{0.245}$$

3) Axial turbine

The centrifugal effect of such a turbine remains slight, and only the proximity comes into play in the expression of C_G. For Z_p/d less than 1, according to the work of Miller and Mann [MIL 44]:

$$C_G = \left(\frac{Z_p}{d} \right)^{-0.12} \qquad \left(\frac{Z_p}{d} > 0.1 \right)$$

The resistance to the flow of the liquid increases much more sharply than the recirculated flowrate decreases, and the power expended increases slightly when the proximity decreases. It tends toward a finite value as Z_P tends toward zero.

4) Radial turbine

Oldshue [OLD 83] proffers curves which we can express as follows:

$$C_G = 0.9 \left[1 - e^{-2.45\frac{Z_I}{d}} \right] \left(\frac{Z_P}{d} \right)^{0.102}$$

Immersion plays a very prominent role because of the centrifugal effect of the radial turbine. This expression is valid only for:

$$1 < \frac{Z_P}{d} < 3$$

Indeed, if Z_P tends toward zero, the resistance to a radial flow does not increase greatly and the radial flowrate does not tend toward zero. It results from this that C_G can never take the value of 0, and must tend toward a finite value (of the order of 0.9).

When Z_P/d surpasses the value of 3, C_G no longer varies.

If we superpose two radial turbines, the C_G are as follows (on condition that Z_S is greater than d):

Upper turbine:

$$C_G = 0.725 \left[1 - e^{-2.3\frac{Z_I}{d}} \right] \left(\frac{Z_S}{d} \right)^{0.242}$$

Lower turbine:

$$C_G = 0.9 \left(\frac{Z_P}{d} \right)^{0.102}$$

For intermediary turbines, we can take $Z_I = 2d$.

5) Bar turbine

The circulation caused by this thruster is slight, and we can accept that $C_G = 1$ for:

$$Z_p > d/8 \qquad \text{and} \qquad Z_I > D/2$$

In practice, these constraints are always satisfied.

In section 1.2.3, we find a table giving the values of N_p and N_Q for the most common types of thrusters.

This table also shows the proximity, the ratio d/D and the range of rotation speeds most commonly used.

1.2.3. *Practical implementation of stirrers in a vat*

Type of thruster	N_P	N_Q	N_Q/N_P	d/D	Z_P	N rev.mn[-1]
Hyper-circulator	0.45	0.65	1.9	<0.7 and >0.4	D/3	20–250
Marine propeller	0.43	0.55	0.63	1/3	D/3	200–1500
Axial turbine	1.4	0.8	0.57	1/4	D/4	200–1500
Radial turbine	6.2	0.7	0.11	1/5	D/4	200–1500
Bar turbine	0.6	~0.1	0.17	1/5	2D/5	500–3000

Table 1.1. *Characteristic numbers for stirrers*

This table corresponds to a vat for which $Z_T/D = 1$ and with 4 chicanes whose breadth is D/10. The values of N_p and N_Q are approximate, as the data given by manufacturers and the published results do not always agree. Only laser velocimetry can deliver satisfactory precision. This technique measures the velocity of particles whose size is of the order of a micron, whose velocity can be said to be identical to that of the liquid.

1.3. Homogenization of a solution

1.3.1. *Objective*

Into a vat filled with a homogeneous liquid mixture, we inject a small quantity of a foreign solution which is soluble in that which is already present in the vat. The aim is to predict the period at the end of which the components of the foreign solution will be uniformly distributed in the liquid filling the vat.

According to Khang and Levenspiel [KHA 76], the homogenization time is given by:

$$\tau = \tau_0 Ln \frac{2}{X}$$

X is the relative fluctuation of the concentration of a given component of the foreign solution in the vat. Industrially, we need to choose $X = 10^{-3}$ (or even 10^{-4}).

These authors injected and measured the concentrations immediately above the thruster, which is to say at its inlet and at a distance $d/2$ from the axis (where d is the diameter of the thruster). The injection and measurement are diametrically opposite in relation to the axis. Such a position for the inlet of the injected solution ensures maximum rapidity of homogenization.

The characteristic time τ_0 is given by:

$$\tau_0 = 1.5 \frac{N_P}{N} \left(\frac{D}{d} \right)^2$$

N_P: power number of the thruster;

N: rotation speed of the thruster: $rev.s.^{-1}$;

D and d: diameters of the vat and of the thruster: m.

EXAMPLE 1.3.–

A vat 3 m in diameter is stirred by a marine propeller 0.8 m in diameter. We inject a small quantity of concentrated acid to modify the pH of the solution. The speed of rotation of the propeller is 4 $rev.s^{-1}$.

$$\tau_0 = 1.5 \times \frac{0.43}{4} \left(\frac{3}{0.8} \right)^2 = 2.3 \text{ s.}$$

$$\tau = 2.3 Ln \left(\frac{2}{10^{-3}} \right) = 17.50 \text{ s.}$$

If the marine propeller is replaced by a hyper-circulator 2 m in diameter and rotating only at 1 revolution per second, we obtain:

$$\tau_0 = 1.5 \times \frac{0.45}{1} \left(\frac{3}{2}\right)^2 = 1.52 \text{ s}$$

$$\tau = 1.52 \times 7.6 = 11.60 \text{ s}.$$

The powers at the shaft necessary in either case are, for the propeller:

$$P = 1000 \times 0.43 \times 4^3 \times 0.8^5 = 9018 \text{ W}$$

For the hyper-circulator:

$$P = 1000 \times 0.45 \times 1 \times 2^5 = 14,400 \text{ W}$$

The hyper-circulator homogenizes the liquid almost twice as quickly, but at the price of almost twice as great an energy expenditure. This is why saber-bladed propellers are reserved for viscous liquids.

1.3.2. *Viscous liquids*

The above correlation remains variable if the viscosity of the liquid mixture is less than 1 Pa.s. According to Oldshue [OLD 83], if we wish to keep constant the homogenization time for viscosities between 1 and 5 Pa.s, we need to increase the rotation speed in the ratio:

$$\frac{N_\mu}{N} = e^{0.3(\mu-1)}; \quad \mu \text{ is the viscosity}$$

Oldshue's results are presented in the form of a curve and a table of values which correspond to the above equation.

1.3.3. *Highly-viscous products*

These products have a viscosity between 5 and 60 Pa.s. They are often pseudoplastics, meaning that their apparent viscosity decreases with the shearing velocity.

Therefore, we can only proceed by extrapolation. In order to do so, we need to have a micropilot and a pilot. If the size of the micropilot is 1 and that of the pilot is k_p, the industrial unit will have the size k_I. The three installations are geometrically similar.

For the micropilot and the pilot, we study the variations in the mixing time with the speed of rotation:

$$\tau_\mu = \tau_\mu(N) \qquad \text{and} \qquad \tau_P = \tau_P(N)$$

By extrapolation, when the similarity ratio passes from the value k_p to the value k_I, we can deduce the value of the parameters from the function:

$$\tau_I = \tau_I(N)$$

We choose the value of N corresponding to a reasonable energy expenditure for the industrial unit.

NOTE.–

It is in our interest to make the stirrers of pilot installations turn slowly if we wish to avoid exorbitant energy expenditures or even impossibilities in the industrial installation.

When this is the case, the process engineer is led to put in place numerous vats stirred in parallel.

Indeed, the power at the shaft of the stirrer increases with $N^3 d^5$ and therefore grows very quickly with the diameter of the thruster.

1.4. Maintenance of a solid in suspension

1.4.1. *Distribution of a divided solid in a vat*

A dispersion of solid in a liquid left alone decants under the influence of gravity. *If it does not decant, then it is a suspension.* To combat the effect of decantation, we maintain circulation in the vat so that any solid having the tendency to be deposited and to accumulate on the bottom must be entrained back to the upper part of the recipient.

For establishing such circulation, only an axial thruster can do the job. Indeed, a radial thruster would require unnecessary energy expenditure. The axial thruster causes the suspension to circulate around the fixed disk. Within that circle, the circulation takes place from top to bottom and, outside of it, from bottom to top. Figure 1.3 illustrates these explanations.

Under the influence of gravity, the solid, which is heavier than the liquid, rises less quickly and descends more quickly. We shall see that this leads to an inequality of the volumetric fractions of solid between the ascending and descending currents.

In addition, at the top and along the horizontal radial trajectory of the liquid, the solid decants. The path of the solid is illustrated by the arrows (1) in Figure 1.3. Consequently, there is no solid at the center of the free surface of the contents of the vat.

Similarly, at the bottom, the solid, whose trajectory is illustrated by the arrows (2), tends to accumulate at the edge on the bottom.

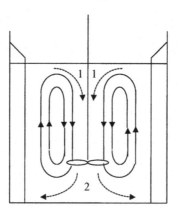

Figure 1.3. *Axial circulation*

1.4.2. *Homogeneity criteria*

The limiting rate of decantation of a cloud of solid particles is $V_{\ell N}$ in a liquid at rest. If the liquid is the site of isotropic turbulence, we can accept that the decantation rate becomes:

$$V_{INT} = 0.4V_{IN}$$

Let $\varnothing_{SR}$ and $\varnothing_{SD}$ represent the volumetric fractions of solid in the suspension on rising and descending and let V_M and V_D be the velocities of the particles with respect to the workshop respectively when rising and descending.

We write that the volume flowrate of the solid is conserved:

$$\varnothing_{SR} V_M = \varnothing_{SD} V_D$$

which can be written:

$$\frac{\varnothing_{SR}}{\dfrac{1}{V_M}} = \frac{\varnothing_{SD}}{\dfrac{1}{V_D}} = \frac{2\bar{\varnothing}_S}{\dfrac{1}{V_M} + \dfrac{1}{V_D}} = \frac{\Delta\varnothing_S}{\dfrac{1}{V_M} - \dfrac{1}{V_D}}$$

with:

$$\Delta\varnothing_S = \varnothing_{SR} - \varnothing_{SD} \quad \text{and} \quad \bar{\varnothing}_S = \frac{1}{2}\left(\varnothing_{SM} + \varnothing_{SD}\right)$$

Thus:

$$\frac{\Delta\varnothing_S}{\bar{\varnothing}_S} = \frac{2\left(V_D - V_M\right)}{V_D + V_M}$$

The decantation velocity is the velocity of displacement of the solid with respect to the immobile liquid. When the liquid is animated with a velocity in an empty bed V_L, the velocity of displacement of the solid with respect to the workshop becomes:

$$V_S = V_{INT} + V_L$$

For an indicative calculation, we shall agree that the surface of the fixed disk is equal to half the horizontal cross-section area of the vat. If the flowrate of clear liquid is Q_L, then:

$$V_L = \frac{2Q_L}{\dfrac{\pi}{4}D^2} = \frac{2Q_L}{A_c}$$

Thus:

$$V_M = \frac{Q_L}{2A_c} - V_{INT}$$

$$V_D = \frac{Q_L}{2A_c} + V_{INT}$$

However, if we accept that the liquid and the solid move at fairly similar speeds, we can write:

$$Q_L = \bar{\varnothing}_L Q_T = \left(1 - \bar{\varnothing}_s\right)Q_T$$

Q_T is the total flowrate recirculated by the stirrer.

By replacing V_M, V_D and Q_L with their expressions, we obtain:

$$\frac{\Delta\varnothing_s}{\bar{\varnothing}_s} = \frac{V_{INT}}{\left(1 - \bar{\varnothing}_s\right)\dfrac{Q_T}{2A_c}}$$

This expression is one way of expressing the heterogeneity of the suspension.

Another way of working is to look for the depth to which, along the axis of the stirrer, the clear liquid is the only thing present.

For this purpose, let us accept that the horizontal radial velocity of the suspension be equal to the velocity V_L calculated above. The time taken by the suspension to cover the distance between the wall of the vat and the axis of the stirrer is:

$$\tau = \frac{D}{2V_L}$$

This gives us the height h_{0S} of liquid freed from the solid:

$$h_{0S} = \tau V_{INT} = \frac{DV_{INT}}{2V_L} = \frac{DV_{INT}A_c}{4Q_L} = \frac{D2V_{INT}A_c}{8\left(1 - \bar{\varnothing}_s\right)Q_T}$$

or indeed:

$$h_{os} = \frac{D}{8} \frac{\Delta \emptyset_S}{\overline{\emptyset}_S}$$

Remember that:

$$Q_T = N_Q N d^3$$

EXAMPLE 1.4.–

Suspension of solid using a marine propeller.

$N = 10$ rev.s^{-1}	$d_P = 2.10^{-4}$ m	$\mu = 10^{-3}$ Pa.s
$D = 3$ m	$\rho_S = 2000$ kg.m^{-3}	$\overline{\emptyset}_S = 0.15$
$D = 0.8$ m	$\rho_L = 1000$ kg.m^{-3}	$N_P = 0.43$

Let us calculate the limiting drop velocity of an isolated particle (see Chapter 4 of [DUR 16] and Chapter 2 of this book):

$$X = \frac{4 \times 9.81 \times 1000 \times \left(2.10^{-4}\right)^3 \times 1000}{3 \times \left(10^{-3}\right)^2} = 104.6$$

$$Y = \frac{13842}{\left(104.6\right)^{1.97}} + \frac{219}{\left(104.6\right)^{1.074}} = 2.93$$

$$V_1 = \left[\frac{4 \times 9.81 \times 10^{-3} \times 1000}{3.10^{-6} \times 2.93}\right]^{1/3} = 0.0165 \text{ m.s}^{-1}$$

Let us now calculate the velocity in a cloud using Richardson and Zaki's formula:

$$Re = \frac{0.0165 \times 2.10^{-4} \times 1000}{10^{-3}} = 3.3$$

$$n = \frac{4.4}{3.3^{0.0982}} = 3.91$$

$$V_{IN} = 0.0165 (1 - 0.15)^{3.91} = 0.00874 \text{ m.s}^{-1}$$

and, in a turbulent regime (see Schwartzberg and Treylal [SCH 68]):

$$V_{INT} = 0.4 \times 0.00874 = 0.0035 \text{ m.s}^{-1}$$

For the marine propeller, the flowrate number is 0.55.

$$Q_T = 0.55 \times 10 \times 0.8^3 = 2.82 \text{ m}^3.\text{s}^{-1}$$

$$A_c = \frac{\pi}{4} \times 3^2 = 7.06 \text{ m}^2$$

$$\frac{\Delta\varnothing_s}{\overline{\varnothing}_s} = \frac{2 \times 0.0035}{(1 - 0.15)\dfrac{2.82}{7.06}} = 0.021$$

$$\phi_s = 0.150 \pm 0.003$$

$$h_{OS} = \frac{3}{8} \times 0.021 = 0.0079 \text{ m}$$

1.4.3. Draft tube

The use of a draft tube is recommended when the ratio of height to diameter of the recipient is greater than 1.3. This device is then able to circulate the suspension of solid from top to bottom and from bottom to top and, therefore, neutralizes the tendency for decantation caused by gravity.

1) Description

The suspension circulates downwards in the tube and upwards in the annular space. This direction of circulation limits the presence of dead zones at the bottom of the vat without the need to overly reduce the space between the base of the tube and the bottom of the vat. This minimizes the pressure drop which the thruster needs to overcome.

The diameter of the tube is 30–50% that of the recipient. The distance from the bottom of the tube to the bottom of the vat is equal to the diameter D_T of the draft tube. To attenuate the vortex effect, the distance from the top of the tube to the free surface of the liquid considered at rest is equal to $D_T/2$. Oldshue [OLD 83, p. 479] gives a diagram of a draft tube. Finally, to reduce the pressure drop, it is recommended to "round off" the bottom of the tube by equipping it with an external torus made with a tube whose diameter is $D_T/4$ (see Figure 1.4).

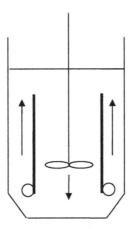

Figure 1.4. *Draft tube*

Circulation is created by a thruster which is often a marine propeller with three blades, but it is also possible to use a hyper-circulator. The velocity at the end of the blade must be no greater than 7 m.s^{-1} for breaking crystals and can reach up to 10 m.s^{-1} if we are dealing with strong particles. To decrease shear at the blade tips, we install a bus 2–3 cm between the thruster and the wall of the tube. The position of the thruster in the tube is indifferent; however, if we want to remedy an untimely decantation having occurred during a stoppage, it is preferable that the propeller should be placed at the bottom.

2) Recirculation flowrate

This flowrate must be sufficient to prevent the progressive accumulation of large particles in the annular space. In this space, the velocity of the liquid must be at least equal to 2 or 2.5 times the limiting velocity of drop of particles whose size corresponds to 10% of refuse from the sieve. This means, for the velocity in the draft tube, that the value must be between 1 and 2 m.s^{-1}. This velocity is limited by the breaking of the crystals or by the wear and tear on the thruster if we are dealing with hard solid particles.

The recirculation flowrate is given by:

$$Q = 0.5 \frac{\pi}{4} d^3 \left(\frac{p}{d} \right) N = V_T \frac{\pi}{4} d^2 \qquad \left(m^3.s^{-1} \right)$$

d: diameter of the thruster: m;

p: step of the propeller: m;

N: frequency of rotation of the thruster: rev.s^{-1}.

The velocity in the tube is then:

$$V_T = 0.5 Nd \left(\frac{P}{d} \right) \qquad \left(m.s^{-1} \right)$$

3) Pressure drop

This is the sum of two terms:

Bottom of the tube (sudden expansion) $1 \times \dfrac{\rho V_T^2}{2}$

Top of the tube (sudden shrinkage) $0.5 \times \dfrac{\rho V_T^2}{2}$

Thus, in total:

$$\Delta P = 1.5 \frac{\rho V_T^2}{2}$$

4) Power at the shaft

This is given by the classic expression:

$$P_a = \frac{Q \Delta P}{\eta}$$

The usual value of the yield η is close to 0.19.

The power number N_P can then be evaluated. Indeed, the power at the shaft is:

$$P_a = \frac{Q\Delta P}{\eta} = \frac{0.5}{0.19} \times \frac{\pi}{4} d^3 N \left(\frac{P}{d}\right) \times 1.5\rho \times (0.5)^2 \times (Nd)^2 \left(\frac{P}{d}\right)^2$$

Thus:

$$P_a = 0.77 N^3 d^5 \left(p/d\right)^3$$

Therefore:

$$N_p = 0.77$$

This value is greater than the value 0.43 for a propeller operating without a draft tube. However, the tube offers the advantage of orienting the circulation along the vertical, which at best neutralizes the effect of decantation due to gravity.

1.5. Dispersion of a gas in a liquid

1.5.1. *Preliminaries*

The dispersion of gases into liquids takes place, as we have seen, with a radial turbine. The gas flowrate must not be greater than a certain limit because, beyond that, the thruster would have difficulty dispersing the bubbles. In that case, we would see flooding.

We can calculate, in turn:

– the power at the shaft of the stirrer;

– the diameter of the bubbles D_B;

– their rate of ascension V_B;

– the volumetric interfacial area a;

– the gas retention R.

The calculation is done by iterations because, at the start, we need to take an approximate value of the retention R.

1.5.2. *Engorgement*

We shall use the expression given by Zlokarnik [ZLO 73]:

$$G < 0.19 \times Fr^{0.75}$$

G is a dimensionless number:

$$G = \frac{Q_G}{Nd^3}$$

Q_G: gaseous flowrate: $m^3.s^{-1}$

Fr is the Froude number:

$$Fr = \frac{N^2 d}{g}$$

g is the acceleration due to gravity ($9.81 \ m.s^{-2}$).

1.5.3. *Power at the shaft*

The power at the shaft is obtained on the basis of the power P_0 in the absence of the gas:

$$P_0 = \rho \, N_p N^3 d^5$$

For a radial turbine, $N_P = 6.2$

We then calculate the coefficient Y:

$$G < 0.03 \qquad Y = 1 - 15 \, G$$

$$G > 0.03 \qquad Y = 0.59 - 1.36 \, G$$

The power at the shaft can then be deduced:

$$P = YP_0$$

1.5.4. *Diameter of the bubbles*

We shall use the expression given by Calderbank [CAL 58]:

$$D_B = K \left(\frac{\sigma^{0.6}}{(P/V)^{0.4} \rho^{0.2}} \right) R^n \left(\frac{\mu_D}{\mu_c} \right)^{0.25}$$

D_B: bubble diameter: m;

σ: surface tension of the liquid: $N.m^{-1}$;

P: power at the shaft of the stirrer: W;

V: volume of the vat: m^3;

ρ: density of the liquid: $kg.m^{-3}$;

R: gas retention;

μ_D and μ_c: dynamic viscosities of the gas and the liquid: Pa.s.

The values of K and N are given by the table below:

Nature of the system	K	n
Gas + electrolyte	2.25	0.4
Gas + water–alcohol solution	1.90	0.65

1.5.5. *Limiting bubble ascension rate*

We shall use Allen's formula [ALL 00]:

$$V_B = \frac{D_B}{4} \left(\frac{\Delta \rho^2 g^2}{\rho \mu_c} \right)^{1/3}$$

V_B: speed of the bubbles: m/s;

$\Delta\rho$: difference between the densities of the liquid and the gas: $kg.m^{-3}$.

In general, this velocity corresponds to the transition zone between the laminar regime and the turbulent regime.

1.5.6. *Volume interfacial area*

The volume interfacial area is given by the expression of Calderbank [CAL 58]:

$$a = 1.44 \sqrt{\frac{V_G}{V_B}} \left[\left(\frac{P}{V} \right)^2 \frac{\rho}{\sigma^3} \right]^{1/5}$$

a: interfacial area: $m^2.m^{-3}$;

V_G: velocity of the gas in an empty bed: $m.s^{-1}$

$$V_G = Q_G / \left(\frac{\pi D}{4} \right)^2$$

1.5.7. *Gas retention*

The gas retention can be calculated if we know a and D_B:

$$R = \frac{aD_B}{6}$$

EXAMPLE 1.5.–

We wish to disperse a gas into a solution of electrolytes with a radial turbine. We estimate that this dispersion will be satisfactory if the maximum rotation speed is 40 $rev.s^{-1}$. The data are as follows:

$\mu_c = 10^{-3}$ Pa.s   d = 0.6vm   D = 3 m

$\sigma = 0.05$ kg/s^2 $\mu_G = 20.10^{-6}$ Pa.s $\rho = 1000$ $kg.m^{-3}$

If we agree that the intensity of turbulence for the covariance is equal to 1, we should have (see [ZLO 73]):

$$40 = 2\pi N \times 2$$

$$N = 3.18 \text{ rev.s}^{-1}$$

$$P_0 = 1000 \times 6.2 \times 3.18^3 \times 0.6^5 = 15,503 \text{ W}$$

$$Fr = \frac{3.18^2 \text{ x} 0.6}{9.81} = 0.62$$

$$G < 0.19 \times 0.62^{0.75} = 0.132$$

$$Q_G < 0.132 \times 3.18 \times 0.6^3 = 0.091 \text{ m}^3.\text{s}^{-1}$$

We shall choose:

$$Q_G = 0.08 \text{ m}^3.\text{s}^{-1}$$

$$V_G = \frac{0.08}{\dfrac{\pi}{4} \times 3^2} = 0.0113 \text{ m.s}^{-1}$$

$$G = \frac{0.08}{3.18 \times 0.6^2} = 0.07$$

$$Y = 0.59 - 0.07 \times 1.36 = 0.495$$

$$\frac{P}{V} = \frac{15503 \times 0.495}{\dfrac{\pi}{4} \times 3^3} = 362 \text{ W.m}^{-3}$$

Suppose that the gas retention is (see [CAL 58]):

$$R = 0.015$$

$$D_B = 2.25 \left[\frac{0.05^{0.06}}{(362)^{0.4} \times 1000^{0.2}} \right] R^{0.4} \left(\frac{20.10^{-6}}{10^{-3}} \right)^{0.25}$$

$$D_B = 0.01682 \ R^{0.4}$$

According to Allen [ALL 00]:

$$V_B = \frac{0.01682 \times R^{0.4}}{4} \left[\frac{10^6 \times 9.81^2}{10^3 \times 10^{-3}} \right]^{1/3}$$

$$V_B = 1.92767 \ R^{0.4}$$

$$a = \frac{1.44}{R^{0.2}} \sqrt{\frac{0.0113}{1.92767}} \left[\frac{362^2 \times 1000}{0.05^3} \right]^{1/5}$$

$$a = 28 / R^{0.2}$$

$$R = \frac{28 \times 0.016826 \times R^{0.2}}{6}$$

Hence:

$$R^{0.8} = 0.07852$$

Thus:

$$R = 0.04156$$

1.6. Dispersion of a liquid insoluble in another liquid (emulsification)

1.6.1. *Liquid–liquid dispersion (see section 1.1.4)*

This operation is performed with a bar turbine if we are dealing with a reaction in a vat. However, if the dispersed phase is unstable and tends toward coalescence and/or decantation, it is necessary to use an axial turbine to create a circulation which brings the large drops and the decanted fraction back to the thruster.

In the plainly-turbulent regime and for a continuous operation, the retention of the dispersed phase is proportional to the corresponding flowrate:

$$R = \frac{Q_D}{Q_D + Q_C}$$

Q_D and Q_C: volumetric flowrates of the dispersed and contiguous phases: $m^3 s^{-1}$.

The interfacial area obtained is given by van Heuven's formula [HEU 70]:

$$a = \frac{6R}{0.047(1+2.6R)} \, 5\sqrt{\left(\frac{P}{V}\right)^2 \frac{\rho}{\sigma^3}\left(\frac{\pi}{4N_P(d/D)^3}\right)^2}$$

A: interfacial area: m^2 per m^3 of dispersion;

P/V: stirring power: $W.m^{-3}$;

ρ: density of the contiguous phase: $kg.m^{-3}$;

N_P: power number of the thruster;

d and D: diameters of the thruster and the vat: m;

σ: interfacial tension between the two liquids: $N.m^{-1}$.

EXAMPLE 1.6.–

$R = 0.40$ $\qquad\qquad$ $P/V = 200 \ W.m^{-3}$ $\qquad\qquad$ $\sigma = 0.03 \ N.m^{-1}$

$\rho = 1000 \ kg.m^{-3}$ $\qquad$ $N_P = 1.4$ (axial turbine) $\qquad$ $d/D = 0.25$

$$a = \frac{6\times 0.4}{0.047(1+2.6+0.4)} \times 5\sqrt{\frac{200^2 \times 10^3}{(0.03)^3}\left(\frac{\pi}{4\times1.4(0.25)^3}\right)^2}$$

$$a = 28,483 \ m^2.m^{-3}$$

1.6.2. *Mean drop diameter*

The mean drop diameter is the harmonic mean, which is to say the diameter of the sphere having the same surface-to-volume ratio as the set of drops.

$$d_{32} = \frac{6R}{a}$$

EXAMPLE 1.7.–

With the above data and results:

$$d_{32} = \frac{6 \times 0.4}{28,483} = 84 \times 10^{-6} \text{ m}$$

1.6.3. *Material transfer coefficient on the side of the contiguous phase*

Calderbank's formula, cited in Uhl and Gray [UHL 60], is written:

$$\beta_c = 0.13 Sc^{-2/3} \left[\frac{(P/V)\mu_C}{\rho^2} \right]$$

β_c: transfer coefficient: $m.s^{-1}$;

S_c: Schmidt number for the contiguous phase;

P/V: stirring power per m^3 of suspension: $W.m^{-3}$;

μ_C: viscosity of the contiguous phase: Pa.s;

ρ_C: density of the contiguous phase: $kg.m^{-3}$.

EXAMPLE 1.8.–

Sc = 300 $\mu_C = 10^{-3}$ Pa.s $\rho = 1000$ $kg.m^{-3}$

P/V = 200 W/m^3

$$\beta_c = \frac{0.13}{(300)^{0.66}} \left[\frac{200 \times 10^{-3}}{10^6} \right]^{1/4}$$

$$\beta_c = 0.61 \times 10^{-4} \ \text{m.s}^{-1}$$

NOTE.–

The expressions giving β_c are numerous in the existing literature. In particular, readers could usefully consult page 244 of Oldshue [OLD 83]. In general, these expressions have only a relative value, and are useful in comparing two installations of different sizes.

1.7. Mixers for pasty products

1.7.1. Description and usage of mixers

In the vat of the mixer, there are two Z-shaped rotors whose axes of rotation are parallel and horizontal. These rotors turn in opposite directions.

Such a mixer must be used when the viscosity of the contiguous phase is greater than 200 Pa.s for a velocity gradient of 5 s^{-1}.

For such a value of the viscosity, the phenomena of diffusion and micromixing no longer occur, and the intensity of segregation remains equal to 100%. The scale of segregation can only be decreased by the influence of shearing, as we shall now see.

1.7.2. Scale of segregation

If the dispersed product occurs in the form of striations, the surface area separating that product and the contiguous product, expressed in relation to the unit volume of the mixture, is:

$$a = \frac{2}{L} \qquad\qquad \left(\text{m}^2.\text{m}^{-3} \right)$$

The segregation thickness L is the minimum orthogonal distance separating two interfaces of the same nature.

Our aim is to disperse a product D, which we introduce in the form of quasi-spherical or cubic domains of size L_o in a contiguous volume of product C with the volumetric fraction $\varnothing_D$.

Let us subject each point of this coarse dispersion to a total shearing force γ for a period of time τ:

$$\gamma = \int_0^\tau \dot{\gamma}\, d\tau = \int_0^\tau \frac{\partial V}{\partial y}\, d\tau = \int_0^\tau \frac{\partial^2 x}{\partial y \partial \tau}\, d\tau = \frac{dx}{dy}$$

Thus:

$$\gamma = \frac{\Delta x}{\Delta y}$$

Δx is the longitudinal displacement, after time τ, of two parallel planes situated the distance Δy apart.

We can show that (see Uhl and Gray, Vol. II, [UHL 66] p. 172) the thickness of segregation has become:

$$L = \frac{L_o}{\gamma \varnothing_D} \frac{\mu_D}{\mu_C}$$

μ_D and μ_C are the viscosities of the products D and C taken separately.

Now suppose that the product D is plastic, meaning that it does not flow unless the shear stress has reached a certain value. The stress transmitted by the matrix C will need to be such that:

$$\mu_C \dot{\gamma} > \text{const.}$$

In summary, the dispersed phase must have as low viscosity as possible.

1.7.3. Power, time and energy

Suppose that the mixing takes place at constant temperature, meaning that the consistency of the products does not vary and that an appropriate cooling device evacuates the heat power dissipated by the viscous friction.

The power per unit volume is proportional to the product of the viscosity by the square of the velocity gradient:

$$P_u \sim \mu \dot{\gamma}^2$$

However, if we express the viscosity using a power law:

$$\mu = K\dot{\gamma}^{n-1}$$

we obtain:

$$P_u \sim K\dot{\gamma}^{n+1}$$

In the expression of the striation thickness, we see the shear effect γ:

$$\gamma = \dot{\gamma}\tau$$

and, for a given rate of shearing γ:

$$P_u \sim K\frac{\gamma^{n+1}}{\tau^{n+1}} = \frac{K'}{\tau^{n+1}}$$

Finally, the energy expended, per unit volume, is:

$$E_u = P_u\tau = \frac{K'}{\tau^n}$$

NOTE.–

The velocity gradient varies depending on the chosen place in the mixer. It is therefore necessary to establish overall circulation whereby each elementary domain experiences the whole range of existing velocity gradients. For this reason, in two-armed mixers, those arms, which rotate around parallel axes, do so at different speeds. Thus, the product moves backward and forward from one arm to the other.

1.7.4. Extrapolation

Large mixers rotate more quickly than smaller ones, which helps to save the energy which would be lost through purely-mechanical friction. The corresponding lost power is simply the no-load power. However, if, for an 8-liter mixer, the no-load power is 300 W, for a mixer whose usable volume is 3.8 m³, the no-load power is only 7.5 kW.

The product is affected only by the difference between the loaded power and the no-load power. This difference is the usable power P*.

The time elapsed for a given degree of shear γ is:

$$\tau = \gamma / \dot{\gamma}$$

However, a constant shear γ means that there is a fixed ratio between the orthogonal gap between two surfaces and their relative longitudinal displacement. If an industrial device is similar to a pilot in the ratio k, the aforementioned orthogonal gaps and longitudinal displacements will also be in the ratio k and, when the slowest axes in each device have completed one turn, the shear obtained is the same. Thus, the shearing forces γ_I and γ_P developed by each device per unit time will be proportional to their respective rotation speeds N_I and N_P:

$$\frac{\dot{\gamma}_I}{\dot{\gamma}_P} = \frac{N_I}{N_P} \quad \left(\text{I for industrial and P for pilot}\right)$$

We know that the unitary power is:

$$P_u = \frac{K'}{\tau^{n+1}} = \left(\frac{K'}{\gamma^{n+1}}\right)\dot{\gamma}^{n+1}$$

Thus, if the shear γ is constant:

$$\frac{P_{uI}}{P_{uP}} = \left[\frac{N_I}{N_P}\right]^{n+1}$$

The total usable power P* is proportional to the volume Ω treated.

Hence:

$$\frac{P_I^*}{P_P} = \frac{\Omega_I}{\Omega_P}\left[\frac{N_I}{N_P}\right]^{n+1} = \left[\frac{D_I}{D_P}\right]^3\left[\frac{N_I}{N_P}\right]^{n+1}$$

NOTE.–

As we have seen, if the dispersed product is plastic, then we must have:

$$\mu_c \dot\gamma = K\gamma_I^n = \text{const.}$$

Thus:

$$N_I = N_P \sqrt[n]{\frac{\text{const.}}{K}}$$

It may therefore happen that we are limited in terms of size for the choice of an industrial mixer if, for mechanical reasons, the manufacturer cannot make a device of the desired size turn fast enough.

1.8. Ribbon mixer (pasty products)

1.8.1. *Usage*

A ribbon mixer should be used when the viscosity of the mixture lies between 50 and 200 Pa.s for a velocity gradient of 5 s^{-1}.

1.8.2. *Description [MER 75]*

This device is formed of a ribbon, whose width is equal to d/10 and which follows a propeller whose step is often square – that is to say, the step p is equal to d. The ribbon is separated from the wall of the vat by a distance e varying between 0.01 and 0.05 d. The minimum value corresponds to greater energy consumption but with a better coefficient of heat transfer at the wall. The ratio d/D therefore lies between 0.90 and 0.98.

The height Z of the vat is equal to its diameter D, and it is closed by two flasks. The space C between each end of the ribbon and the flask which closes the vat at those ends is equal to 0.03 d.

The device often includes a second ribbon adjacent to the axis and whose propeller diameter is d/3. Its step is equal to that of the outer ribbon, but it is oriented in the opposite direction.

1.8.3. *Power consumed*

The power number is obtained using Käppel's [KAP 70] relation.

$$N_P = \frac{60}{Re}\left(\frac{p}{d}\right)^{-0.5} n^{0.8}\left(\frac{e}{d}\right)^{-.03}$$

p: step of the outside ribbon: m;

d: diameter of the outside ribbon: m;

e: distance from the outside ribbon to the wall: m;

Re is the Reynolds number:

$$Re = \frac{\rho N d^2}{\mu}$$

EXAMPLE 1.9.–

$d = 2m$ $\qquad$ $\mu = 100$ Pa.s $\qquad$ $\rho = 1000$ kg.m^{-3}

$N = 0.7$ rev.s^{-1} $\qquad$ $e = 0.02$ d $\qquad$ $P = d$

$n = 2$

$$Re = \frac{1000 \times 0.7 \times 2^2}{100} = 28$$

$$N_P = \frac{60}{28} \times 1 \times 2^{0.8} \times (0.02)^{-0.3}$$

$$N_P = 12$$

$$P_u = 1000 \times 12 \times 0.7^3 \times 2^5 = 131.712 \text{ W}$$

$$P_u = 132 \text{ kW}$$

1.8.4. *Mixing time*

As the products at hand are often pseudoplastics and, generally speaking, non-Newtonian, it is wise to proceed by extrapolation, as we did for mixers.

However, Mersmann *et al.* [MER 96] put forward the relation:

$$\tau^2 = 5.8 \times 10^5 \frac{\mu D^3}{P_u}$$

EXAMPLE 1.10.–

$\mu = 100$ Pa.s $D = 2.04$ m $P_u = 132 \times 10^3$ W

$$\tau^2 = \frac{5.8 \times 10^5 \times 100 \times (2.04)^3}{132 \times 10^3} = 3730 \text{ s}^2$$

$\tau = 61 \text{ s} = 1 \text{ min}$

1.8.5. *Heat transfer at the wall*

The transfer coefficient is of the order of 20 W.$^{-2\circ}$C^{-1}.

Dispersion and Dissolution of Powders

2.1. General points about powders and crystals

2.1.1. *Properties of powders relative to dispersion/dissolution*

The amorphous form dissolves more quickly than the crystalline form does but, when stored, it can gradually transform into crystalline form.

The anhydrous form absorbs moisture much more readily than the hydrated form, which may present disadvantages during storage (clumping). However, the anhydrous form dissolves more quickly in the digestive tract.

When a crystal is polymorphous and if we want a suspension to be stable, we must use the less-soluble form, because it swells, to the detriment of the more-soluble form.

The size of the crystals is important. A very fine powder dissolves more quickly than if it is made of coarser particles.

Pinpoint injection of a suspension of crystals is easier if the crystals are platelets or, better yet, are equidimensional, than if the crystals have the form of stems.

Certain difficulties may manifest themselves, depending on the powder we wish to disperse. The powder may be:

– predisposed to sedimentation: a small amount of gelling agent will cause thixotropic behavior in the suspension;

– crumbling: the dust released must be sucked out to maintain the hygiene of the workshop or else the operation will be carried out in an enclosed vat;

– foaming: perform in an enclosed, pressurized vat with slow stirring;

– composed of aggregates: turbulence and particularly shearing are recommended;

– floating on the surface: place the thruster of the stirrer in the vicinity of the surface;

– predisposed to the formation of a gangue (certain starches). In such cases, the product needs to be heated and/or pre-mixed with a highly-soluble powder.

Finally, dispersion and therefore hydration may take a long time: flours (in baking or pastry-making), pectins (the time required may be up to 24 hours), granulates (formed of aggregates).

On the other hand, certain powders are very quick to dissolve, such as instant powders (which are highly porous), resulting from lyophilization or else flash drying (drying by rapid depressurization).

Let us cite a few examples taken from the food industry, pertaining to dispersion/dissolution:

– lyophilized coffee dissolves in only a few seconds, because it is highly porous, almost to a molecular scale;

– fine table salt (d_p of around 100 μm) obtained by milling, dissolves in around ten seconds;

– atomized coffee comes in microspherules and dissolves in a few tenths of a second;

– atomized milk contains lipids which are insoluble in water and are hydrophobic. The other components disperse and dissolve in a less than a minute;

– crêpe flour swells in water in an hour. The granules are then dispersed;

– coarse-grained salt is not designed to disperse or dissolve, but instead to give an intense salty flavor.

2.1.2. Concept of dispersibility

The International Dairy Federation published a standard by which to evaluate the dispersibility of powdered milk (IDF 87 standard).

Using a palette, we disperse a mass P of powder in liquid by very gentle stirring. After a period of time τ, we filter the suspension obtained on a metal grid whose aperture is d. Particles and aggregates whose size is less than d constitute the passthrough, which gives us the mass p by desiccation of the suspension in a vat. The dispersibility is then (for τ and d):

$$D = \frac{p}{P}$$

2.1.3. Conditions for good dispersion

Regardless of the means by which a powder is obtained (milling, jet micronization, drying by atomization, precipitation by chemical reaction), aggregates may exist when brought into contact with a liquid. Indeed, the surface energy of a fine powder ($d_p < 0.1$ μm) is high and the van der Waals forces of attraction are significant, especially if the particles are non-polar. The resulting aggregates must be easy to destroy.

The operation of dispersion in water must ensure complete contact between the liquid and the surface of the particles. It must also prevent sedimentation under the influence of gravity. All of this, though, depends on the properties of the powder.

1) The wetting of the solid surface must be at least partial (existence of a contact angle) and, if possible, perfect or even better, more than perfect (i.e. a hydrophilic surface).

2) The true density of the particles must not differ from that of water by more than 15%. More generally, the particle size must be such that the descent rate in water is less than 10^{-4} m.s^{-1} so that the dispersion of the particles does not give way to sedimentation.

$$\frac{g\Delta\rho d_p^2}{18\mu} < 10^{-4} \text{ m.s}^{-1}$$

3) The energy of agitation to be employed for a dispersion can vary in a ratio of 1 to 10 in view of the diversity of the characteristics of powders but, especially, the variability, from case to case, of the ratio:

$$\frac{\text{mass of powder}}{\text{mass of liquid}}$$

EXAMPLE 2.1.–

$$g = 9.81 \text{ m.s}^{-2} \qquad \Delta\rho = 100 \text{ kg.m}^{-3} \qquad \mu = 10^{-3} \text{ Pa.s}$$

Thus:

$$d_p < 43 \times 10^{-6} \text{ m}$$

This size is of a typical order of magnitude for the powders to be dispersed.

2.2. Physics of wetting

2.2.1. *Surface energy and contact angle*

The surface tension of a liquid is half the traction force which needs to be exerted on a thin film of liquid of unitary width to prevent it retracting. The surface tension is measured in $N.m^{-1}$ but it would be equivalent to say that it is measured in $J.m^{-2}$ and, in this case, we would be dealing with a surface energy.

Meanwhile, for liquids, we use the notion of surface tension, for solids we use the notion of surface energy, because in the case of a solid, the notion of a thin retracting film is meaningless.

Now suppose that a drop of liquid is laid on a solid surface (see Figure 2.1) and remains there in a stable state.

Let us write the static equilibrium of the point O where the three interfaces come together, characterized by the following surface energies:

γ_{sv} : between the solid and the vapor of the liquid at the saturating pressure at the temperature of the system.

γ_{SL} : between solid and liquid.

γ_{LV} : between the liquid and its saturating vapor.

Let us consider on the surface of the solid the forces that are exerted on point O.

$$\gamma_{SV} = \gamma_{SL} + \gamma_{LV} \cos\theta \qquad [2.1]$$

This is Young's relation and θ is the contact angle.

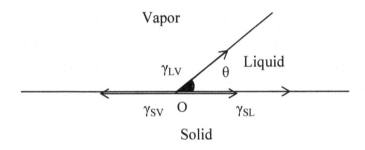

Figure 2.1. *Contact angle*

2.2.2. *Helmholtz energy and enthalpy of immersion*

Consider a powder whose total surface area is A. Supposing this is possible, let us immerse that powder in a liquid, the variation in surface energy of the solid is:

$$A\left(\gamma_{SL} - \gamma_{SV}\right) = -A\gamma_{LV} \cos\theta$$

This is the Helmholtz energy that has been lent to the powder.

By calorimetry (with a Dewar), we can measure the enthalpy Δh_i gained by the powder–solvent system on immersion. For a total surface area A, the energy is $A \Delta h_i$.

For example, with the immersion of titanium oxide in water, we obtain:

$$\Delta h_i = 0.325 \text{ J.m}^{-2}$$

2.2.3. Energy of adherence. Dupré's relation [DUP 69]

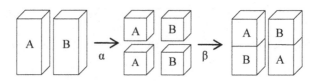

Figure 2.2. *Dupré's relation*

Consider two solid phases whose vapor tensions we can discount, and therefore that we shall consider to be placed in a vacuum. These two solids have the form of prisms whose section is a half-unit. During step α, we section the prisms in parallel to their base, using the Helmholtz energy $(\gamma_A + \gamma_B)$. During step β, we permute the lower half-prisms and make them solid with the upper prisms. We apply the energy of adherence β^*. The result is the interfacial energy γ_{AB}, which is written:

$$\gamma_{AB} = \gamma_A + \gamma_B + \beta*$$

Let us stress the fact that the *seizure energy β to be applied to the system to separate the surfaces A and B is* $-\beta^*$. Thus, Dupré's equation is written:

$$\gamma_{AB} = \gamma_A + \gamma_B - \beta \qquad [2.2]$$

Armed with this relation, suppose that A is a solid and B a liquid, all surrounded by the vapor of the liquid at the saturation pressure corresponding to the temperature of the system.

Dupré's relation [DUP 69] becomes:

$$\gamma_{SL} = \gamma_{SV} + \gamma_{LV} - \beta \qquad [2.3]$$

Eliminate $\gamma_{SV} - \gamma_{SL}$ between equations [2.1] and [2.3]. The seizure energy β is then written:

$$\beta = \gamma_{LV}(1 + \cos\theta) \qquad [2.4]$$

In the above, γ_{LV} and γ_{SV} are always positive.

2.2.4. *Energy of cohesion*

If, in equation [2.2], we make $A \equiv B$, then $\gamma_{AB} = 0$, because there is no interfacial energy within a phase. Thus, we obtain:

$$\beta_c = 2\gamma_A$$

β_c is the energy of cohesion of the phase A. For a liquid in contact with its saturating vapor:

$$\beta_c = 2\gamma_{LV}$$

Thus, the energy of cohesion of water is:

$$\beta_c = 2 \times 0.078 = 0.156 \text{ J.m}^{-2} \quad (\text{or N.m}^{-1})$$

2.2.5. *Wetting, spreading and Helmholtz energy*

Depending on the value of the adherence, we may encounter five situations.

1) $\beta < 0$ meaning that $\gamma_{SV} + \gamma_{LV} - \gamma_{SL} < 0$

The replacement of the SL interface by a thin gaseous with two faces SV and LV corresponds to a decrease in Helmholtz energy. In other words, spontaneously, the solid and the liquid repel one another, e.g. if their electrical charges are of the same sign. If the liquid is water, the solid surface is said to be hydrophobic, and this situation is desired in the operation of flotation.

2) $\beta = 0$; according to relation [2.4]:

$$\cos \theta = -1 \text{ and } \theta = \pi$$

The drop is deposited on the solid at a single point. Strictly speaking, there is no wetting of the solid. The wetting is *zero* and is therefore impossible.

3) $\beta_c > \beta > 0$

If we detach the liquid from the surface of the solid, rupture occurs at the interface rather than within the body of the liquid. The wetting is *imperfect*.

For a degree of advancement ΔA of the front of the drop in parallel to the solid surface, the variation in Helmholtz energy is:

$$\Delta F = \left(\gamma_{SL} + \gamma_{LV}\cos\theta - \gamma_{SV}\right)\Delta A$$

Yet according to Young's relation, the term in parentheses is equal to zero. Hence, $\Delta F = 0$, and the equilibrium is *indifferent*.

However, mechanical stirring will be necessary to completely wet the particles.

4) $\beta = \beta_c$

If we detach the liquid from the solid, rupture will occur just as readily either at the interface or within the liquid.

In view of equation [2.4]:

$$2\gamma_{LV} = \gamma_{LV}\left(1 + \cos\theta\right) \text{ and hence } \cos\theta = 1 \text{ and } \theta = 0$$

The wetting is said to be *perfect*. The equilibrium is *indifferent*. A certain degree of stirring is still necessary.

5) $\beta > \beta_c$

The value of $\cos\theta$ can be no greater than 1, and we always have $\theta = 0$, in which case the wetting is said to be *more than perfect*. If we detach the liquid from the solid surface, the liquid breaks far from this surface, leaving a liquid film with strong adherence on the solid.

If, on the surface ΔA, we replace the interface SV with a thin film with 2 faces LV and SL, the variation in Helmholtz energy is:

$$\Delta F = \Delta A \left(\gamma_{LV} + \gamma_{SL} - \gamma_{SV}\right)$$

However, because the wetting is more than perfect, we have $-\beta_c + \beta > 0$:

$$\beta_c = 2\gamma_{LV} \text{ and, in view of equation [2.2], } \beta = -\gamma_{SL} + \gamma_{SV} + \gamma_{LV}$$

Thus:

$$-\gamma_{LV} - \gamma_{SL} + \gamma_{SV} > 0$$

Consequently:

$$\Delta F < 0$$

The spreading of the liquid is *spontaneous*, and no agitation at all, or else very weak stirring, is sufficient to bring about dispersion. Such is the case of soluble particles. If the liquid is water, the solid surface is said to be hydrophilic, just like cotton cellulose of the same name, or like the surface of a crystal in contact with its parent liquor in the under-saturated state.

We can write:

$$\beta = \beta_c + \Delta\gamma = 2\gamma_{LV} + \Delta\gamma \text{ where } \Delta\gamma > 0$$

Let us feed this expression into Dupré's equation [2.2]. We obtain:

$$\Delta\gamma = \gamma_{SV} - \gamma_{SL} - \gamma_{LV} > 0$$

Δγ is the "driving force" of spontaneous wetting. This surface energy is called the "*spreading energy*", or rather, the "energy available for spreading".

The parameter $\Delta\gamma$, which is *positive*, cannot be found directly by experimentation. In reality, we can measure the rate of advancement of the thin liquid film at the surface of the solid. This rate is probably less than the rate of advancement of a liquid film at the surface of another liquid, which is of the order of a few centimeters per second.

2.2.6. *Wettability (practical aspect)*

The IDF 87 standard from the International Dairy Federation can be used to quantify the idea of wettability.

We uniformly distribute a normalized mass of powder on the surface of the liquid and note the necessary time τ:

– for the particles to submerge below the surface;

– and for the particles having remained on the surface take on the typical "wet" appearance.

The measurement thus performed is not particularly precise, but can be used to monitor the fluctuations of manufacturing.

2.2.7. Measurement of the total surface energy of a liquid

This measurement is taken using a vertical rectangular plate whose long side is horizontal and has the length L. The plate is immersed in the liquid and then raised out of it. As it comes out, it entrains a liquid film whose horizontal length is L. The force needed to spread that film is:

$$F = 2L\gamma + p$$

γ: surface energy: $J.m^{-2}$.

We can also use a horizontal ring whose perimeter is P.

In this case, we have:

$$F = 2P\gamma + p$$

The weight of the plate or the ring is p. There are excellent tensiometers that are commercially available.

2.2.8. Measurement of the contact angle on a powder

To begin with, we create a "tablet" at a pressure of 100–200 bar. That tablet is brought into contact with water on its lower surface, so that it becomes saturated with liquid.

A drop of liquid is deposited on the "tablet". Its height h is independent of the mass of the drop. As the tablet is saturated with liquid, it does not absorb the liquid from the drop, which therefore remains stable.

Kossen *et al.* [KOS 65] showed that the contact angle θ is such that:

$$\cos\theta = 1 - \sqrt{\frac{Bh^2}{3(1-\varepsilon)\left(1-\frac{Bh^2}{2}\right)}} \quad \text{where } B = \frac{\rho_L g}{2\gamma_{LV}} \qquad [2.5]$$

γ_{LV}: surface tension of the liquid in the presence of its vapor.

The above relation is valid for $0 < \theta < 90°$.

When $\theta > 90°$, Hansford *et al.* [HAN 80] mention an extension of relation [2.5].

$$\cos\theta = -1 + \sqrt{\frac{2}{3(1-\varepsilon)}\left(\frac{2}{Bh^2}-1\right)} \qquad [2.6]$$

The drop height h is measured with a cathetometer.

This method measures the mean contact angle on the faces of a type of crystal without it being necessary to operate on a crystal of notable size.

Numerous contact angles are given in the publications of Lerk *et al.* [LER 76, LER 77].

2.2.9. *Wetting agents*

Wetting agents are designed to ensure a bond between water and a hydrophobic surface. We distinguish:

1) Ionic agents, for which:

i) the lipophilic part is the anion (anionic agents):

– alkyl-sulfates,

– alkyl-sulfonates,

– triethanolamine soaps;

ii) the lipophilic part is the cation (cationic agents). Such agents include salts of amines or of quaternary ammonium.

2) Amphoteric agents, may be:

– anionic in a basic medium;

– cationc in an acidic medium.

3) Non-ionic agents (unaffected by pH):

– hydroxyl groups and ether bridges ensure solubility in water.

The length of the hydrophobic part allows us to regulate the ratio of hydrophilic to lipophilic behavior.

The hydrophobic parts (lipophiles) are:

– fatty acids;

– fatty alcohols;

– alkyl-phenols;

– polyalcohol esters.

NOTES.–

These agents ensure more than perfect wetting of the solid surface by water.

2.2.10. *Dispersive energy and polarization energy*

There are always instantaneous distorsions of the electron clouds, which cause an instantaneous polarization in their immediate vicinity. These dispersions (which occur in turn at a very high frequency) thus engender forces of attraction between electron clouds. This is at the heart of the forces of dispersion.

Forces of polarization are due to permanent electrical polarization of the molecules, which are therefore attracted to one another, often by way of hydrogen bonds.

The above remains valid for the surface of a medium. Hence, we divide the surface energy into two terms: γ^d for dispersion and γ^p for polarization.

$$\gamma = \gamma^d + \gamma^p$$

Wu [WU 71] performed the same division for the energy of adherence.

$$\beta = \beta^d + \beta^p$$

and, between two media 1 and 2, wrote:

$$\beta^d = \frac{4\gamma_1^d \gamma_2^d}{\gamma_1^d + \gamma_2^d} \quad \text{and} \quad \beta^p = \frac{4\gamma_1^p \gamma_2^p}{\gamma_1^p + \gamma_2^p}$$

2.2.11. Practical measurements

We exploit the fact that, for solid paraffins, $\gamma_{SV}^p = 0$ and that $\gamma_{SV}^d = 0.0255\ \text{N.m}^{-1}$

For a liquid in contact with paraffin, we measure the total γ_{LV} and the contact angle θ, which gives us $\beta = \beta^d$, because $\beta^p = 0$ (indeed, $\gamma_{SV}^p = 0$). From β^d, we deduce γ_{LV}^d.

Knowing the total surface energy (surface tension), we deduce:

$$\gamma_L^p = \gamma_{LV} - \gamma_{LV}^d$$

Zografi et al. [ZOG 76] applied this method to water and methylene iodide. They found that the latter liquid was 100% dispersive and that water was 32% dispersive $\left(\gamma_{LV}^d / \gamma_{LV} = 0.32\right)$.

When we know the characteristics of these two liquids, it becomes possible to calculate those for any given solid by applying the following relation to each of the two liquids.

$$-\beta = \gamma_{LV}\left(1 + \cos\theta\right) = 4\left[\left(\gamma_{LV}^d \gamma_{SV}^d\right)^{1/2}\left(\gamma_{LV}^p \gamma_{SV}^p\right)^{1/2}\right] \qquad [2.7]$$

NOTE.–

When we look at equation [2.7], we can see that the energy of adherence is maximal when:

$$\gamma_{SV}^p = \gamma_{LV}^p \text{ and the common value is high.}$$

On the other hand, the energy of adherence is minimal if:

$$\text{or } \left.\begin{array}{l} \gamma_{SV}^p \ll \gamma_{LV}^p \text{ and } \gamma_{SV}^d \gg \gamma_{LV}^d \\ \gamma_{SV}^p \gg \gamma_{LV}^p \text{ and } \gamma_{SV}^d \ll \gamma_{LV}^d \end{array}\right\} \quad \text{when } \gamma_{SV} \text{ and } \gamma_{LV} \text{ are constants}$$

Put differently, $-\beta$ is minimal if the solid (or liquid) is polar and if the liquid (or solid) is dispersive.

More simply, we can state that adherence is mediocre if the two media are of opposite characteristics. Thus, water, which is essentially polar, does not easily wet fatty substances, which are essentially dispersive.

NOTE.–

The expression given by Wu [WU 71] for the energy of adherence shows that β is always positive and may very well be greater than $\beta_c = 2\gamma_{LV}$ (be greater in absolute value).

The negative value β corresponds to the absence of contact. In this case, it is illusory to speak of the energy of cohesion.

2.3. Practice of dispersion – equipment

2.3.1. *Dispersion procedures (food industry)*

Depending on the characteristics of the system, we may wish to:

1) Destroy any aggregates:

– in a vat, use a palette stirrer rather than a propeller. We then see the formation of eddies behind the palettes, and when the aggregates hit the blades, they are destroyed by the shock;

– to create a colloidal dispersion, we could use a mill composed of two adjoining disks (the rotor and stator), fed by the axle. In this case, shearing is intense;

– the tremors caused by ultrasound (or any other mechanical vibration) will destabilize the aggregates;

– we can add a repulsion agent or a tensioactive agent (to facilitate wetting of the elements in the aggregates);

– we can build in an external recirculation circuit through a centrifugal pump (where the shearing force is always significant, which destroys the aggregates).

2) Rid the mixture of the gaseous phase (aggregates contain an occluded gaseous phase and are therefore unable to penetrate into the liquid).

We need to:

– de-aerate the fine powder ($d_p < 100$ μm) to obtain a regular flow of it;

– put a dust-removal mechanism in place (an aspiration fan and filter) if the powder is "crumbling", i.e. if it releases pollutant dust;

– ensure the position of the thruster of the stirrer at the bottom of the vat will prevent the incorporation of air into the vortex, whilst combatting any sedimentation.

3) Modify the system's physical properties:

– exploit the temperature (and consequently the viscosity and surface tension of the liquid);

– add a small quantity of gelling agent to increase the viscosity of the liquid and thereby prevent the sedimentation of the heavy particles;

– perform dry premixing with a highly soluble powder.

2.3.2. *Destruction of aggregates*

If we wish to disperse n particles in a volume V of liquid, the liquid volume devoted to each particle will be:

$$\frac{V}{n} = \frac{1}{c}$$

c: concentration: particles.m^{-3}.

The mean distance between two particles will, when dispersion has been carried out, be:

$$l_e = \frac{1}{c^{1/3}}$$

However, the particles are introduced in the form of aggregates each containing n_{pa} particles.

If the position of the aggregate is position zero, the particles need to be removed by:

$$l_e, \, l_e \times 2, \, l_e \times 3 \ldots \ldots l_e \times (n_{pa} - 1)$$

This corresponds to a total displacement, in light of the fact that the sum of the first n numbers is n(n − 1)/2:

$$l_e (n_{pa} - 1)(n_{pa} - 2)/2$$

The displacement must take place in all three spatial directions, i.e.:

$$3l_e (n_{pa} - 1)(n_{pa} - 2)/2$$

To separate the particles, the appropriate means is to subject the suspension to a velocity gradient dv/dz.

Let d_p represent the diameter of the particles. The relative velocity v of a particle in relation to the neighboring particle is:

$$v = \left(\frac{dv}{dz}\right) d_p$$

After a period of time τ_e, the elementary separation distance, l_e, traveled is:

$$l_e = \tau_e \left(\frac{dv}{dz}\right) d_p$$

Thus, we have the time τ_e:

$$\tau_e = \frac{l_e}{(dv/dz)d_p}$$

and the time taken for the destruction of the aggregates will be:

$$\tau_{ea} = \frac{3l_e(n_{pa}-1)}{d_p(dv/dz)}$$

By a calculation which we shall not detail here, it is possible to show that the usable energy expended for the dispersion, per unit volume, is very small:

$$W_u = \frac{3\mu}{2}\left(\frac{dv}{dz}\right)^2\frac{d_p}{l_e}(n_{pa}-1)(n_{pa}-2)$$

This is the energy expenditure due to the viscous forces. As regards the work required to separate the particles, it is even more negligible:

$$W_s = \frac{Ad_p}{24l_e^3}Ln\left(\frac{l_e}{D_0}\right) \text{ and } W_s = \frac{E_s}{\tau_{ea}}$$

A: Hamaker constant: 10^{-18} J;

D_{in}: initial distance of the surfaces of the two particles: $D_{in} = 50\times10^{-9}$ m.

EXAMPLE 2.2.–

$$\mu = 5\times10^{-3} \text{ Pa.s} \qquad \frac{dv}{dz} = 5 \text{ s}^{-1} \qquad n_{pa} = 20$$

$$d_p = 10 \text{ µm} \qquad n_{pa} = 20 \qquad \ell_e = 50 \text{ µm}$$

The time necessary for the dispersion is:

$$\tau_{ea} = \frac{3\times5\times(20-1)}{5} = 57s \#1 \text{ mn}$$

The usable viscous power will be:

$$W_u = \frac{3 \times 5 \times 10^{-3} \times 25 \times 19 \times 18}{2 \times 5} = 12.82 \text{ W.m}^{-3}$$

The separation power will be:

$$W_s = \frac{10^{-18} \times 10 \times 10^{-6}}{24 \times \left(50 \times 10^{-6}\right)^3 \times 57} \times \ln\left(\frac{50 \times 10^{-6}}{50 \times 10^{-9}}\right) = 0.0063 \text{ W.m}^{-3} \# 0$$

However, the usual power of a stirrer is around 1 kW.m^{-3}, *which is greater than the usable power by a factor of around 100.*

2.3.3. *Dispersion in the pharmaceutical industry*

In the case of a medicine, its dispersion by the gastric liquids does not depend on the choice of a device, but rather on the way in which the active principle is presented for bio-assimilation. We distinguish between;

1) prior crystallization of a eutectic formed of the active principle and a carrier adjuvant that is soluble in water. Indeed, the grains of the active principle of small size (μm) are immediately dispersed during the dissolution of the carrier.

2) solid solution of the carrier and the active principle. On the dissolution of the carrier, the medicine is dispersed at the molecular scale. The solid solution is therefore preferable to the eutectic. The carrier may be a polymer obtained by evaporation of solvent. Glycol polyethylenes are soluble in water and act as a protective colloid which prevents clumping of the particles of medicine.

These two procedures can be used for the manufacture of a tablet or the filling of a capsule. The active principle is then released on digestion.

2.3.4. *Principles of action of devices for mechanical dispersion*

Shearing is used in the laminar or mixed regime. Thus, in a micro-bearing milling machine, there are 600 times more shearing points than in a disk miller (let τ_{shear} represent the shear stress):

$d_{bearing}$	$\tau_{shear.}$
1 mm	350 bar
3 mm	17,000 bar

In a paint mill (or, strictly speaking, a disperser), the ratio of the distance between the cylinders to their diameter is:

$$\frac{e}{D} = \frac{15\,\mu m}{0.3\,m} = 5 \times 10^{-5}$$

The stability of a dispersion in relation to gravity increases with the viscosity of the liquid. On the other hand, a low viscosity is a positive factor for the flocculation of the suspension.

NOTE.–

To obtain an emulsion of one liquid in another, it is necessary to operate by shocks (toothed disk). The distribution between the contiguous phase and the dispersed phase depends on the respective volumes of the two liquids. However, it is possible to create an emulsion of oil in water with only 4% water per volume.

2.3.5. *Choice of devices*

An important parameter is the viscosity of the continuous phase:

1) $0.01\ Pa.s < \mu < 1\ Pa.s$ $(1cp = 10^{-3}\ Pa.s)$:

– devices causing shock by sudden changes in direction of the fluid threads (particularly for emulsions);

– gap of a few μm between two surfaces slipping over one another (shearing);

– ultrasound, which causes very great shearing.

2) μ # a few tens of Pa.s:

– bearings situated between two coaxial cylinders whose speed of relative rotation may be around 1000 rev.min^{-1}. The bearings range from 1 to 3 mm in diameter;

– toothed disk turning at 1500 rev.mn^{-1}. The shearing, which is significant, depends on the peripheral velocity;

– cylinder dispersers;

– pasters-kneaders (e.g. a baker's kneading machine) and mixers for dispersions ("sols") of plastic material (incorporation of dye) with Z-shaped blades. These (charge) devices consume a power between 50 and 300 kW and have a usable volume between 60 liters and 1.5 m^3.

NOTE.–

Powders are generally obtained in the humid phase and the aggregates form when the substance is dried. It may therefore be of interest to directly disperse the cake from humid filtration.

2.3.6. *Paint and plastic material industries*

To disperse pigments and other adjuvants in an oil, a solvent or indeed a plasticizer, the difficulty needing to be overcome is how to destroy the aggregates of particles. We obtain this result by subjecting the dispersion to a significant velocity gradient known as shearing.

Thus, the paint disperser is composed of three rollers turning more or less quickly in the direction of progression of the product. The different velocities present a twofold interest:

– ensuring the progression of the dispersion from the slow roller to the fast roller;

– causing shearing between the cylinders.

Figure 2.3. *Roller disperser*

During their contact, the distance of relative slip of two contiguous cylinders is:

$$\Delta L = T_c \times \Delta v$$

T_c: contact time on $1/30^{th}$ of the circumference: s

$$T_c = \frac{T_{rev}}{30} = \frac{1}{30N}$$

Δv: difference between the peripheral velocities

$$\Delta v = \pi D \Delta N = \pi D \left(\frac{\Delta N}{N} \right) N$$

Hence:

$$\Delta L = \frac{\pi D}{30} \left(\frac{\Delta N}{N} \right)$$

The shear deformation between the two cylinders, if e is the distance between them, is:

$$\frac{\Delta L}{e} = \frac{\pi D}{30e} \left(\frac{\Delta N}{N} \right)$$

The velocity gradient will then be:

$$\left(\frac{dv}{dz} \right) = \frac{\Delta L}{T_c} \times \frac{1}{e} = \frac{\pi D \Delta N}{e}$$

EXAMPLE 2.3.–

$$\Delta N = 0.13 \text{ rev.s}^{-1} \qquad D = 0.3 \text{ m} \qquad e = 0.5 \times 10^{-3} \text{ m}$$

$$\left(\frac{dv}{dz} \right) = \frac{\pi \times 0.3 \times 0.13}{0.5 \times 10^{-3}} = 245 \text{ s}^{-1}$$

This value is around 50 times greater than the velocity gradients in a vat equipped with a classic stirrer.

2.3.7. *Other industrial dispersion processes*

We draw a distinction between continuous-flow dispersion and dispersion in a vat. Indeed, the time devoted to the operation may be extremely long if the powder needs to swell (pectins, flours) and run to several hours, which, in principle, rules out a continuous-flow operation which, for its part, should last no more than a few seconds.

1) Continuous-flow dispersion. We could use:

– a line mixer such as a longitudinal flux divider marketed by Sulzer on condition that the particles are not prone to clumping;

– the introduction of a predetermined flowrate of powder (dosing device) at the neck of a venturi where the liquid is circulating;

– a regular feed of powder must facilitate a correct flow continuously. However, if clotting occurs, it can block the dosing system and break the electrical circuit.

2) Dispersion in the vat. We simply use a recipient with a palette stirrer.

2.4. Dissolution of a small crystal and dissolution of a powder

2.4.1. *Affinity for water (hygroscopicity)*

Polar crystalline surfaces attract water molecules. If they carry positive surface charges, they attract the oxygen atoms of the water molecule. If they carry negative charges, they attract the hydrogen atoms in the same molecule.

The anhydrous form of a chemical species obviously has more affinity for water than a hydrated form.

When the wetting of a surface by a liquid is said to be "more than perfect", not only is the contact angle zero, but the liquid does not remain in the form of a static wet stain, instead tending to spread indefinitely over the wetted surface. For this to happen, it is sufficient for the energy of adherence of the liquid to the solid to be more than twice the surface tension of the liquid.

Naturally, the wetting of a hygroscopic surface is "more than perfect".

Hygroscopicity is attributable to the predominance of the hygroscopic faces of a crystal. Indeed, the affinity for water is not the same for all faces.

When a hygroscopic powder is stored in too humid an environment, points of saline solution are established between the particles and, if the humidity decreases, these points solidify, causing a gain in the mass of the stored powder. This phenomenon could be called "massification". It is for this reason that certain powders are stored in an inert atmosphere. Furthermore, the absence of oxygen can be beneficial. Thus, hoppers, kegs and even sacks are designed for this purpose.

Adjuvants and excipients, if they are hydrophobic, slow down the disaggregation of the tablets and, conversely, encourage such disaggregation if that are hydrophilic and, more importantly, water-soluble. Such is the case, in particular, with disintegrants.

Gelatin, starch and microcrystalline cellulose exhibit a hysteresis in their humidity absorption curves. The explanation for this phenomenon lies in the fact that the retained water is found on three different types of sites:

– monomolecular layer connected to the surface of the product;

– multimolecular layers situated on the previous layer;

– internal humidity of the biological material.

The humidity in the monomolecular layer is subject to two types of forces:

– surface bond forces;

– forces of diffusion toward the interior of the biological material.

During desorption, there is no force to extract the humidity from the biological material, as long as not all the humidity has disappeared from the surface. This is the reason for the hysteresis:

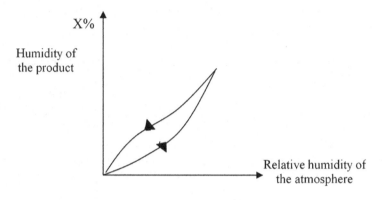

Figure 2.4. *Hysteresis of adsorption*

In conclusion, the affinity for water is conditioned by:

– the penetration of the liquid into tablets and granulate materials before they disaggregate;

– the dissolution of the particles resulting from that disaggregation;

– the adherence of the coating liquid for tablets and pharmaceutic pills.

2.4.2. *Description of the dissolution of a crystal*

Very generally, the faces of a crystal are never ideally planar. They contain defects which may be punctual (lack of or excess of molecules at a given point) or linear, meaning changes in level (steps due to dislocations). The number of these defects typically ranges between 10^5 per m² of surface and 10^8 m⁻².

The liquid in which the concentration of solute is lower than the concentration at equilibrium attracts the molecules with these defects, because the Helmholtz energy (dispersible) of these molecules is greater than the Helmholtz energy of molecules which are wisely inserted into the crystalline lattice (both at the surface and within the crystal).

Thus, hollows form on the faces, as do "notches" on the edges, i.e. at all sites which are energetically favorable. Let us specify that these sites are often due to a defect in the underlying structure, and these imperfections have been preserved during the "reconstruction" of the superficial layer (whose structure is not always identical to the structure deep within the crystal – hence the idea of "reconstruction").

Yet for undersaturations that are much less than 1%, the formation of hollows is sufficiently slow to be the dominant phenomenon for dissolution. When the undersaturation increases notably, the diffusion of the molecules through the limiting film surrounding the crystal becomes dominant. It is this hypothesis which has been adopted in calculating the dissolution times with undersaturation equal to:

$$\sigma_D = \frac{c^* - c_\infty}{c^*}$$

c^*: concentration at equilibrium: $kmol.m^{-3}$ or $kg.m^{-3}$;

c_∞: concentration within the liquid: $kmol.m^{-3}$ or $kg.m^{-3}$.

Note that the undersaturation thus defined must necessarily be less than or equal to 1:

$$0 < \sigma_D < 1$$

The kinetics of formation of hollows around the defects does not appear to have been studied. However, there is a correspondence which exists between the oversaturation having led to the crystal's growth and the rate of dissolution of that crystal when it has moderate undersaturation.

2.4.3. *Rate of attack of the surface and oversaturation of crystallization*

When a crystal has been obtained with great oversaturation, the surface density of imperfections is maximal. Conversely, in a crystal obtained slowly on the basis of slight oversaturation, the density of imperfections is minimal.

As we have seen, the rate of dissolution of a crystal is an increasing function of the density of imperfections. The rate of extraction of the molecules grows perhaps with the logarithm of that density.

The overall dissolution rate, then, is a combination of the extraction rate and the rate of diffusion of the molecules across the limiting film surrounding the crystal.

In summary, the faster a crystal has grown, the more quickly it will dissolve (for limited subsaturation).

2.4.4. *Theory of the diffusion layer*

We suppose that the crystal to be dissolved is surrounded by a "diffusion layer". The concentration on contact with the crystal is c* (corresponding to equilibrium) and the concentration within the liquid is c_∞.

1) The convective layer has a thickness δ given by:

$$\delta = \frac{d}{2 + 0.6\,Re^{1/2}\,Sc^{1/3}}$$

d: dimension of the crystal: m

Sc: Schmidt number

$$Sc = \frac{\mu}{\rho D}$$

μ: viscosity of the liquid: Pa.s

D: diffusivity of the solute in the liquid: $m^2.s^{-1}$

ρ: density of the liquid: $kg.m^{-3}$

Re: Reynolds number

$$Re = \frac{ud\rho}{\mu}$$

u: velocity of the crystal in relation to the liquid: $m.s^{-1}$

In a fluidized bed:

$$Re = \frac{Ud\rho}{\varepsilon\mu}$$

ε: porosity of the fluidized bed;

U: velocity of the liquid in an empty bed: m.s^{-1}.

For a stirred suspension:

$$Re = d^{4/3}W^{1/3}\rho\mu^{-1}$$

W: mechanical power per unit mass of the suspension: W.kg^{-1}.

The molar concentration (kmol.m^{-3}) at the internal surface of the diffusion layer is c_0. On the external surface, it is equal to c_∞.

The molar flux density across the diffusion layer is:

$$\varphi = \frac{D}{\delta}(c * - c_\infty) \qquad (kmol.m^{-2}.s^{-1})$$

The molar flux density traversing the diffusion layer is:

$$\varphi = \frac{D}{\delta}(c * - c_\infty) = \frac{D}{\delta}c * N_A \sigma_D$$

This corresponds to a medium decrease rate.

$$R_D = \Omega\varphi = \frac{D\Omega c * \sigma_D}{\delta} \qquad (m.s^{-1})$$

σ_s: under-saturation;

c*: concentration of the solution at equilibrium: kmol.m^{-3};

Ω: molecular volume: m^3.molecule^{-1};

N_A: Avogadro's number: 6.02×10^{26} molecules.kmol^{-1}.

2.4.5. *Time for dissolution of a single crystal in an infinite volume of liquid*

Suppose that the temperature of the system is constant and that the mass of solvent is sufficient so its concentration does not vary when the crystal dissolves.

The saturation of the solution is:

$$\Delta c = \sigma c^* < 0$$

c^*: concentration at equilibrium: $kmol.m^{-3}$

σ: saturation rate (here, negative)

$$-1 < \sigma < 0$$

The material transfer coefficient is, for a crystal supposed to be spheroidal (equidimensional) with average diameter d_p

$$k = \frac{2D}{d_p} \quad \left(\text{supposing that Re} = 0\right) \qquad (m.s^{-1})$$

The variation of mass of the crystal per unit time is:

$$\frac{2D}{d_p} S\Delta cM = S\frac{d(d_p)}{2d\tau}\rho_c < 0$$

M: molar mass of the species under consideration: $kg.kmol^{-1}$;

S: surface of the crystal: m^2;

ρ_c: density of the crystal: $kg.m^{-3}$.

Let us integrate in order to find the time at the end of which the crystal will be dissolved:

$$\tau = \frac{\rho_c(0-d_p^2)}{DM\Delta c \times 8}$$

EXAMPLE 2.4.–

$$\rho_c = 1500 \text{ kg.m}^{-3} \qquad\qquad c^* = 2 \text{ kmol.m}^{-3}$$

$$d_p = 10^{-3} \text{ m} \qquad\qquad M = 50 \text{ kg.kmol}^{-1}$$

$$D = 10^{-8} \text{ m}^2.\text{s}^{-1}$$

$$\Delta c = 2 \text{ kmol.m}^{-3}$$

$$\tau = \frac{10^{-6} \times 1500}{10^{-8} \times 50 \times 2 \times 8} = 187 \text{ s} = 3 \text{ min } 7 \text{ s}$$

2.4.6. *Dissolution of a crystal in a limited volume of liquid*

In order for the liquid to be able to absorb all of the solid in the crystal, we must have:

$$(c^* - c_0) MV > \frac{\rho_c d_p^3 \pi}{6} \qquad\qquad [2.8]$$

c^*: concentration at saturation: kmol.m^{-3};

c_0: initial concentration: kmol.m^{-3};

M: molar mass of the solute: kg.kmol^{-1};

V: volume of the liquid: m^3;

ρ_c: density of the crystal: kg.m^{-3};

d_p: dimension of the crystal: m.

This relation enables us to calculate the minimum volume of liquid V to obtain complete dissolution.

When the dimension of the crystal has passed from its initial value d_{p0} to the value d_p, the concentration has moved from c_0 to c:

$$VM\ (c - c_0) = \frac{\pi}{6}\ (d_{p0}^3 - d_p^3)$$

[2.9]

If the material transfer coefficient is $2D/d_p$, the mass transferred from the unit surface of the crystal is written as follows, over the time period $d\tau$:

$$\frac{2D}{d_p}(c* - c) = -\frac{\rho_c}{2M}\frac{d\ (d_p)}{d\tau}$$

[2.10]

Let us eliminate c between equations [2.9] and [2.10] and posit:

$$\frac{\pi \rho_c}{6VM}\ (d_p^*)^3 = c* - c_0 - \frac{\pi \rho_c}{6VM} d_{p0}^3$$

We obtain:

$$d\tau = -\frac{3V}{2\pi D}\frac{d_p d(d_p)}{\left(d_p^3 + (d_p^*)^3\right)}$$

[2.11]

Let us break this fraction down into simple elements:

$$\frac{6ax}{x^3 + a^3} = -\frac{2}{x + a} + \frac{2x - a}{x^2 - ax + a^2} + \frac{3a}{x^2 - ax + a^2}$$

The discriminant of the second-degree formula is negative. The integration of equation [2.11] gives us:

$$\tau = -\frac{V}{4\pi D}\left[-2Ln\left(\frac{d_p^*}{d_{p0} + d_p^*}\right) + Ln\left(\frac{(d_p^*)^2}{d_{p0}^2 - d_{p0}d_p^* + (d_p^*)^2}\right) + 2\sqrt{3}Arctg\left(\frac{-d_p^*}{d_p^*\sqrt{3}}\right) - 2\sqrt{3}Arctg\left(\frac{2d_{p0} - d_p^*}{d_p^*\sqrt{3}}\right)\right]$$

EXAMPLE 2.5.–

$$M = 50 \text{ kg.kmol}^{-1} \qquad \rho_c = 1500 \text{ kg.m}^{-3} \qquad D = 10^{-8} \text{ m}^2.\text{s}^{-1}$$

$$c^* = 2 \text{ kmol.m}^{-3} \qquad d_{p0} = 10^{-3} \text{ m} \qquad c_0 = 0$$

$$V_{min} = \frac{1500 \times \pi \times 10^{-9}}{50 \times 2 \times 6} = 0.07 \times 10^{-7} \text{ m}^3$$

We choose : $V = 10^{-7} \text{ m}^3 \Delta V_{min}$

$$d_p^* = \left[\frac{6 \times 10^{-7} \times 2 \times 50}{\pi \times 1\,500} - 10^{-9} \right]^{1/3}$$

$$d_p^* = 0.002224 \text{ m}$$

$$\tau = -\frac{10^{-7}}{4\pi \times 10^{-8} \times 2.224.10^{-3}} \left[\begin{array}{l} 2\mathrm{Ln}\left(\dfrac{3.224}{2.224}\right) - \mathrm{Ln}\left(\dfrac{1-2.224+2.224^2}{2.224^2}\right) + \\[2mm] 2\sqrt{3}\,\mathrm{Arctg}\left(-1/\sqrt{3}\right) - 2\sqrt{3}\,\mathrm{Arctg}\left(\dfrac{2-2.224}{2.224\sqrt{3}}\right) \end{array} \right]$$

$$\tau = 211 \text{ s} = 3 \text{ mn} \, 31 \text{ s}$$

This dissolution time is obviously a little longer than if the volume of liquid were infinite (we found 187 s).

2.5. Continuous-flow dissolution of a suspension

2.5.1. Continuous-flow dissolver (Figure 2.5)

The dissolution of crystals is generally five times faster than the reverse operation, i.e. crystallization. Knowing that the length of stay of the slurry in a crystallizer is generally between 1 and 4 hours, the solvent's length of stay of the solvent in a dissolver to obtain a saturated solution will vary between a quarter of an hour and an hour.

A dissolver includes a highly-stirred dispersion/dissolution zone (with a palette thruster) and a peripheral zone where the solution obtained is clarified.

The volume of the shaded cylindrical part is the usable volume for dispersion and dissolution. Its value is:

$$V_D = Q_S \tau$$

V_D: usable volume: m^3;

Qs: flowrate of solution: m^3/s;

τ: dissolution time: s.

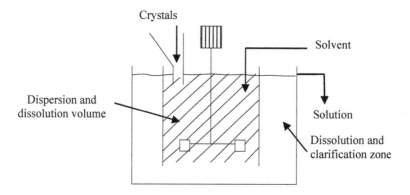

Figure 2.5. *Continuous-flow dissolver*

2.6. Specific cases

2.6.1. *Dissolution of capsules, tablets and pills*

1) Capsules

The content of the capsule decreases exponentially after a latency period. The capsule may have been filled with powder, whether compressed or otherwise.

Capsules of hard gelatine become breakable at low humidity and become sticky when the humidity is high. They must therefore be filled in carefully controlled conditions $(30 < H\% < 50)$ both for the container and for the content.

2) Tablets

There is a relation of proportionality (see below) between the depth of penetration of the liquid in the tablet and the square root as the time (see section 2.6.2). A tablet will dissolve more easily when:

– the permeability of the tablet is not too low;

– the initial powder is fine;

– the particles of the powder have the amorphous form, but the amorphous particles can gradually transform into crystals;

– the anhydrous form is present, but we need to protect the tablets from the humidity in the atmosphere;

– if the solid is present in several crystalline forms, we must use the most soluble of those forms.

3) Dissolution of pills

Pharmaceutical pills are tablets coated in a protective layer. Thus, this layer needs to be dissolved first, but that dissolution may be very quick.

2.6.2. *Theory of penetration of a liquid into a porous medium (tablet or granulated powder)*

We cannot expect a liquid to penetrate into a porous medium unless the make up of the grains of that medium is more than perfectly wetted by the liquid. This is expressed by the relation:

$$\gamma_{SV} - (\gamma_{SL} + \gamma_{LV}) = \Delta\gamma > 0 \quad \text{(more than perfect wetting)}$$

The capillary ascension theory gives us the force acting on the liquid in a pore with radius r.

$$F = \pi r^2 \Delta P = 2\pi r \Delta\gamma$$

Thus:

$$\Delta P = 2\Delta\gamma / r$$

Poiseuille's law applied to the liquid cylinder whose radius is r, filling the pore over the length l, is expressed by:

$$\frac{dl}{d\tau} = \frac{\Delta P r^2}{8\mu l} = \frac{r\Delta\gamma}{4\mu l}$$

We integrate:

$$l^2 = \frac{r\Delta\gamma\tau}{2\mu} \qquad\qquad [2.12]$$

τ: time: s;

μ: viscosity of the liquid: Pa.s;

$\Delta\gamma$: excess surface energy: N.m^{-1}.

If the porous solid is composed of spheroidal particles whose diameter is d_p, stacked up with the porosity ε, the mean pore radius is given by [DUR 99]:

$$r = \frac{\varepsilon d_p}{3(1-\varepsilon)}$$

For porosities ε between 0.35 and 0.6, we can accept that the tortuosity of the pores is near to $\sqrt{2}$. If h is the depth of penetration of the liquid measured perpendicularly to the surface of the porous substance, we have:

$$l = h\sqrt{2}$$

Let us replace r and l with their expressions in equation [2.12]:

$$2h^2 = \frac{\Delta\gamma\varepsilon d_p}{6\mu(1-\varepsilon)}\tau \qquad\qquad [2.13]$$

or indeed:

$$h = \left[\frac{\Delta\gamma\varepsilon d_p}{12\mu(1-\varepsilon)} \right]^{1/2} \sqrt{\tau} = K\sqrt{\tau} \qquad\qquad [2.14]$$

The depth of penetration of the liquid is proportional to the square root of the elapsed time.

The volume occupied by the liquid in a cylinder of height h and section of unitary area is simply:

$$\Omega_L = \varepsilon h$$

In equation [2.13], all the parameters are known apart from $\Delta\gamma$, which it is therefore possible to calculate.

2.6.3. *Influence of pressing on the dissolution of a tablet*

This influence is expressed by a fourfold action:

1) Rearrangement of the particles and decreasing of the porosity of the tablet, which slows the penetration of the liquid.

2) Plastic deformation and then fragmentation of the particles lead to an increase in surface area per unit volume of the whole, which will be favorable for dissolution.

3) These mechanical actions cause defects in the crystals, which will encourage dissolution later on.

4) However, under the influence of pressure, the particles interpenetrate with one another, by what could be called "cold welding". This effect is negative in relation to the dissolution rate, because it decreases the surface area available for contact with the liquid.

NOTE.–

The solidification of the active principle in a metastable form may occur. The rate of dissolution of such a form is quicker than that of a crystallized form. If the metastable form is a glass where the intermolecular bonds are as weak as in a liquid, this facilitates dissolution. If the metastable form is crystallized, the advantage is lesser, but present nonetheless.

A tablet will dissolve more easily if:

– the permeability of the tablet is not too low;

– the initial powder is fine;

– the particles of the powder have an amorphous form, but these amorphous particles may slowly transform into crystals;

– the anhydrous form is present, but we need to protect the tablets against the humidity present in the atmosphere;

– if the solid occurs in various crystalline forms, we need to use that which is the more soluble.

Mixture of Divided Solids: Choice of Mixing Devices

3.1. Criteria for evaluating the homogeneity of a mixture

3.1.1. *Introduction*

The mixing of divided solids is used in numerous industries. For example, we could cite: ceramics, plastics, grease, detergents, glass, pharmaceuticals, human and animal nutrition, etc.

In this chapter, we shall not discuss mixing in fluidized beds, or the state of mixing when a silo or hopper is emptied. Furthermore, mixing in a fluidized bed is to be avoided when the ratio between the densities of the two components is greater than 2.5.

3.1.2. *Mean, variance and variation coefficient*

Consider a divided solid (DS) made up of particles whose composition may vary. Let us take n samples of that DS, and let x_i be the composition of the sample with index i.

1) Mean

The mean or, more specifically, the arithmetic mean of the values obtained, is:

$$\overline{x} = \frac{\sum\limits_{i=1}^{n} x_i}{n}$$

2) Variance and standard deviation

The variance s is given by:

$$s^2 = \frac{\sum\limits_{i=1}^{n}(x_i - \overline{x})^2}{n}$$

Certain authors use the term "variance" for the square s^2.

The standard deviation is:

$$\sigma = \sqrt{\frac{\sum\limits_{i=1}^{n}(x_i - \overline{x})^2}{n-1}}$$

Note that the standard deviation thus evaluated is infinite if the number of samples is equal to 1.

3) Variation coefficient (which would be more aptly called the "relative variance")

The variation coefficient is defined as follows, for a given component.

$$C_v = \frac{s}{x}$$

s: variance for the component in question;

x: concentration of the component in the mixture.

3.1.3. *Binary mixtures, noteworthy identities*

Stange [STA 54] gives the classic equations linking the variance of the sum, the product and the quotient of two variables. They are recapped here:

Consider two variables a and b:

$$s^2(a \pm b) = s^2(a) + s^2(b)$$

Note the absence of the minus sign on the right-hand side here.

$$s^2(a.b) = \overline{a}^2 s^2(b) + \overline{b}^2 s^2(a)$$

$$s^2\left(\frac{a}{b}\right) = \left(\frac{\overline{a}}{\overline{b}}\right)^2 \left(\frac{s^2(a)}{\overline{a}^2} + \frac{s^2(b)}{\overline{b}^2}\right)$$

From this, we deduce, where k = const.:

$$s^2(k\,a) = k^2 s^2(a) \quad \text{and} \quad s^2\left(\frac{k}{a}\right) = \frac{k^2}{\overline{a}^4} s^2(a)$$

3.1.4. *Variance of a totally separated (unmixed) binary system. Intensity of segregation*

An unmixed binary system can be considered to contain n_p particles of (P) and n_q particles of (Q). For this mixture, the variance is:

$$s_o^2 = \frac{\sum_1^n (\overline{x} - x_i)^2}{n_p + n_q}$$

The mean value of $\bar{x}$ is as follows, for the same heterogeneous mixture, with the particles being drafted one by one:

$$\bar{x} = \frac{n_p \times 1 + n_q \times 1}{n_p + n_q} = 1$$

For exhaustive drafting, the proportion of particles (P) drafted is:

$$x_p = \frac{n_p}{n_p + n_q} = p; \text{similarly}: x_q = \frac{n_q}{n_p + n_q} = q \text{ where } p + q = 1 \text{ and } n = n_p + n_q$$

Hence:

$$s_1^2 = \frac{\sum_1^{n_p}(1-p)^2 + \sum_1^{n_q}(1-q)^2}{n_p + n_q} = \frac{n_p(1-p)^2 + n_q(1-q)^2}{n_p + n_q} = p(1-p)^2 + q(1-q)^2$$

$$s_1^2 = p(1 - 2p + p^2) + (1-p)p^2 = p(1-p)$$

s_1^2 is the variance of sampling of the unmixed (completely separated) system.

The intensity of segregation is defined by [DAN 53] for the mixture of A and B:

$$I = \frac{s^2}{\overline{a.b}} = \frac{s^2}{x(1-x)} = \frac{s^2}{s_1^2}$$

The intensity I varies, during mixing, from 1 to a limit lying somewhere between 1 and zero ([LAC 76], part I).

3.1.5. Reference variances

Consider a DS composed of particles (P) and (Q) in the proportion p and q where:

$$p + q = 1$$

Suppose we have a homogeneous mixture.

Let us take a series of m samples each containing n particles. For each sample, analysis gives us the composition p_j.

[LAC 43] showed that the variance is written:

$$s^2 = \frac{\sum\limits_{j=1}^{m}\left(p_j - \bar{p}\right)^2}{m} = \frac{pq}{n}$$

This relation corresponds to an imperfectly-homogeneous mixture.

If, on the other hand, the mixture is completely separated, as we have just seen:

$$s^2 = p(1-p) = pq$$

3.1.6. *Reference sample too large for the counting of the particles*

Suppose that the number m of particles of the DS is too large to be accurately determined. In this situation, we replace the number of particles in a sample with the mass M of that sample.

The particles of each species are generally not uniform in size, and therefore we need to bring into play the distribution of those particles in terms of diameter or (which is equivalent) in terms of unitary mass.

Poole *et al.* [POO 64] gave a useful method for obtaining the variance s of the distribution of a component in the set of samples. Consider x and y:

$$s^2 = \frac{xy}{M}\left[y(\Sigma fm)_x + x(\Sigma fm)_y\right] \qquad [3.1]$$

x and y: weighting fractions of the component X and of the component Y;

M: mass of a sample: kg;

f: weighting fraction of one of the components in the fraction whose mean mass is m;

m: mass of the particles of diameter d: kg.

[VAL 67] gives a detailed precise example of the calculation of s using Poole *et al.*'s formula [POO 64].

This formula shows that $\log_{10}s$ is a linear function (whose slope is –0.5) of $\log_{10}M$.

3.1.7. *Measuring the variance [SOM 74, SOM 76, SOM 82]*

$$s^2 = s_{tot}^2 = s_M^2 + s_z^2 + \left(1 - \frac{g_1}{G}\right)s_{syst} = s_E^2 + \left(1 - \frac{g_1}{G}\right)s_{syst}$$

s_M^2 : variance due to inaccurate measurements;

s_z^2 : variance obtained if the mixture were ideally and randomly mixed;

$s_E = s_M^2 + s_z^2$ does not vary during the course of the mixing.

g_1 is the maximum mass of an individual particle and G is the mass we have chosen for each sample. The ratio g_1/G is generally much less than 1.

[SOM 74, SOM 75] verified that:

$$s_z^2 = \frac{P(1-P)}{n}$$

n: number of particles in a sample;

s_Z expresses the random variations of concentrations due to heterogeneity in le mixer.

s_{syst} is the systematic variance. This is the true criterion which needs to be chosen to evaluate the quality of the mixture. This variance expresses the systematic variations in concentration depending on the location in the mixer where the sample is taken.

s_{syst} is independent of the size of the samples.

If only a few samples are chosen, the confidence limits expand, and there is a risk of acceptance of a mixture which, in reality, is not right.

If we improve the accuracy of the measurement, s_E can be lowered, so that the acceptable upper bounds for s_{syst} and s become equal (see the authors' Figure 5). In other words:

> *The more accurate the measurements, the fewer samples are needed.*

3.1.8. Influence on the variance of the number and size of samples

1) The number of samples

[BOU 67] brought into play the autocorrelation coefficient (see section 3.2.1) in the following analytical form:

$$\rho_k = a^k \quad -1 \le a \le 1$$

For a random ideal mixture, $a = \rho = 0$.

For a completely separated "mixture", $a = \rho = 1$.

Let n be *the number of samples* taken in the mixture.

[BOU 67] showed that the variance of the result of the measurements is expressed by:

$$s_n^2 = \frac{s_1^2}{n^{1-a}}$$

He demonstrated and experimentally verified that:

$$\ln s_n = \ln s_1 - \left(\frac{1-a}{2}\right) \ln n$$

s_1 is the variance for a single sample.

By experimentation, it is possible to obtain the value of $\left(\dfrac{1-a}{2}\right)$, which ranges between 0.5 and 0.27.

[STA 67] carried out a study in the same domain and confirmed the results found by [BOU 67].

2) Let us now examine the influence of *the unitary mass* of the samples.

Y- or Λ-shaped mixers rotate in their plane (Figure 2 p. 20 of [LAC 76a]).

The distribution of the composition of the samples tends toward a normal distribution as mixing continues.

The intensity of segregation I decreases as we increase the size m of the samples, and this decrease is all the quicker when the number of revolutions is high, i.e. when the energy expended is high.

m: mass of a sample;

E: energy expended for mixing (number of revolutions performed by the mixer).

When the slope $\dfrac{d \log_{10} I}{d \log_{10} m}$ is slight, this indicates that the segregation is long range.

An almost-horizontal straight line represents a totally-separated mixture and a slope equal to –1 represents a random ideal mixture.

[VAL 67], in his Figure 2, shows that, in line with Stange's results, there is no longer a linear relation between $\log_{10} s$ and $\log_{10} m$ when the distribution of the measured value is not a normal distribution.

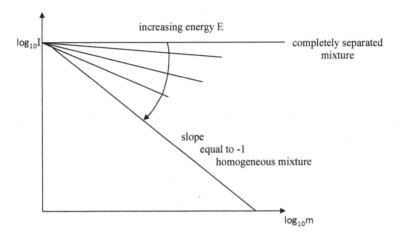

Figure 3.1. *Segregation index as a function of Z and of N*

3.1.9. *Variance reduction ratio*

$$VRR = \frac{s_0^2}{s_1^2}$$

Equation 19 in [WEI 95] is written:

$$\frac{1}{RRV} = 2 \int_0^\infty I(\tau)\rho(\tau)d\tau + \frac{s_1^2}{s_0}$$

$\rho(r)$: autocorrelation function (see section 3.2.1)

They obtained the first term by integration of the equation established by [FAN 79], based on Planck's equation [PLA 17].

3.1.10. *Angle of repose*

[CHO 79] tell us that the natural of repose angle is obtained by measuring the height and the radius at the base of the cone obtained when the DS is poured onto a planar surface through a hole.

That angle varies as an inverse function of the flowability, and thus can be used to classify divided solids on the basis of their flowability.

3.1.11. *Diameter distribution desirable for the active ingredient*

[STA 54] gave the expression of the variance of a binary random mixture as a function of the particle size distributions (PSDs) of the active and excipient ingredients.

[JOH 72] took up Stange's demonstration, employing Poisson's law (see [BOG 06] p. 56) but above all, he used the result found by Poole *et al.* [POO 64]. Finally, the expression of the variance of the distribution of number of active particles per sample is given by:

$$s^2 = \frac{xy}{M}\left[y(\Sigma fm)_x + x(\Sigma fm)_y\right] \tag{3.2}$$

The quality parameter adopted by [JOH 72] is the active variation coefficient (relative variance).

$$C_v = \frac{s}{x} \text{ or indeed, in percentage } C_v = \frac{100s}{x}$$

[JOH 72] hypothesizes, for his calculations, that the size distribution of the active ingredient is normal. He shows that, for a C_v of 10%:

– supposing that all the particles of the active ingredient are of the same size, we can calculate the limiting particle diameter as a function of x;

– step by step, we can obtain the size distributions as a function of x. The "limiting distribution" corresponds to a C_v of 1%. Johnson bases his work here on that of [LAN 72].

3.2. Autocorrelation function

3.2.1. *Definition*

The autocorrelation function aims to check whether there is a correspondence (matching) between the concentrations at two different places in a DS, with those two places being a distance r apart. The variance cannot, on its own, reflect this correspondence.

[DAN 52] was the first to define the autocorrelation coefficient, in his equation 5. The coefficient reflects the matching over short- and medium distances.

[LAC 76] looked at the issue in detail and, of three possible formulations, chose to use the following:

$$\rho(r) = \frac{\sum_{i=1}^{n-1}(x_i - \overline{x})(x_{i+r} - \overline{x})}{\left[\sum_{i=1}^{n-1}(x_i - \overline{x})^2\right]^{1/2} \times \left[\sum_{i+2}(x_{i+r} - \overline{x})^2\right]^{1/2}}$$

Figure 3.2 illustrates the evolution of $\rho(r)$ as a function of the distance r.

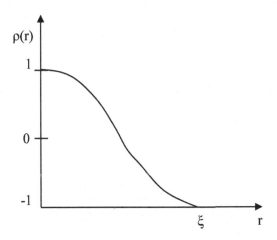

Figure 3.2. *The function ρ(r)*

The scale of segregation is the area situated beneath the curve $\rho(r)$.

$$S = \int_0^\infty \rho(r)\,dr = \int_0^\xi \rho(r)\,dr$$

The curve $\rho(r)$ is called the correlogram.

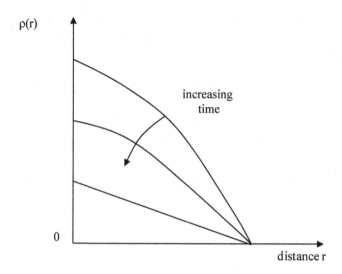

Figure 3.3. *Evolution of autocorrelation*

The autocorrelation tends toward zero, according to [GYE 99], when the mixing time increases.

Indeed, mixing reduces large clumps of particles ([LAC 76a]).

NOTE.–

[DAN 53] puts forward practical elements regarding the autocorrelation coefficient $\rho(r)$, and proposes methods by which to measure $\rho(r)$.

3.3. Acceptance of the quality of a mixture

3.3.1. *Probability of an event and probability density*

Suppose that we know the frequency distribution of the probability P of appearance of the value x of the property we are measuring:

$$\varphi(x) = \frac{dP}{dx}$$

The probability that the value x will fall within the interval [a, b] is then:

$$P(a < x < b) = \int_a^b \varphi(x)dx = P(b) - P(a)$$

The probability P is equal to the surface between x = a and x = b beneath the curve $\varphi(x)$.

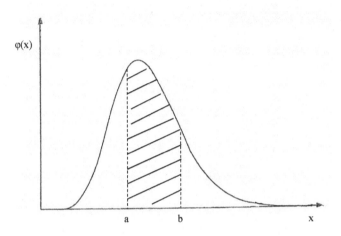

Figure 3.4. *Frequency distribution of the probability of occurrence of the value x*

3.3.2. *Central limit theorem [BOG 06; p. 221]*

The central limit theorem is stated thus:

The sum of n random variables X_1, ...X_n divided by $\sqrt{n}$ tends toward a reduced normal distribution as n increases indefinitely, on condition that the X_i are centered reduced random variables which are independent and identically distributed.

The characteristic function of a reduced normal distribution is:

$$\varphi(t) = e^{-\frac{t^2}{2}}$$

Remember that the probability density function $\varphi(x)$ for the so-called Gauss–Laplace normal law is:

$$\varphi(x) = \frac{1}{\sigma\sqrt{2\pi}} \exp\left[-\frac{1}{2}\left(\frac{(x-\mu)}{\sigma}\right)^2\right]$$

3.3.3. *Estimation by confidence interval*

The aim here is to give a range of values which has a known probability of containing the unknown value.

If we know the distribution of the values of x_i, we can determine a *confidence interval* [a, b] such that:

$$P(a \leq x_i \leq b) = 1 - \alpha \qquad\qquad [3.3]$$

with:

$$P_1(x_i < a) = \alpha_1 \quad \text{and} \quad P_2(x_i > b) = \alpha_2$$

and:

$$\alpha_1 + \alpha_2 = \alpha$$

α is the probability that the double inequality [3.3] will not be satisfied. The most commonly used values are $\alpha = 0.01$ and $\alpha = 0.05$. In addition, it is reasonable to stick to the middle of the interval of probabilities P_1 and P_2 so that the bounds a and b correspond to the quantities $\frac{\alpha}{2}$ and $1 - \frac{\alpha}{2}$ in the distribution.

Now consider a set of n values x_1, ..., x_n, and we shall – in line with the central limit theorem – accept that they obey a Gaussian normal probability law with mean μ and standard deviation σ. It is possible to show (though we shall not do so here) that the variance s^2 can be estimated as lying within a specific interval, with the probability $P = P_2 - P_1$, if we have fixed a value for the probabilities P_1 and P_2. Thus, for the probability $1 - \alpha$:

$$P\left[\chi^2_{\alpha/2} \leq \frac{(n-1)s^2}{\sigma^2} \leq \chi^2_{1-\alpha/2}\right] = 1 - \alpha \qquad [3.4]$$

The square s^2 is distributed in accordance with the so-called χ^2 law (pronounced "chi-squared"). This distribution is known, and tables give the value of the χ^2 corresponding to the values $\alpha/2$ and $1 - \alpha/2$ for a given value of the parameter n, which is called the number of degrees of freedom and which, remember, is the number of elements in a sample.

The inequalities [3.3] can be written thus, to know the standard deviation σ:

$$\frac{(n-1)s^2}{\chi^2_{1-\alpha/2}} \leq \sigma^2 \leq \frac{(n-1)s^2}{\chi^2_{\alpha/2}}$$

NOTE.–

Thus, we have obtained a confidence interval for the standard deviation of the distribution of the measurements $x_1, ..., x_n$. To do so, we accepted the hypothesis that *this distribution was normal*. Note that the mean value of the measurements is:

$$\mu = \frac{\sum\limits_{i=1}^{n} x_i}{n}$$

This distribution is perfectly well known, and therefore we can obtain a confidence interval for the value μ accepted for s. If the parameter n is less than or equal to 5, the central limit theorem no longer applies, and we must use Student's distribution to calculate the confidence interval for μ. We shall not linger over this point here (see [BOG 06, p. 297]).

The variance s^2 is different from the experimental estimation of the standard deviation, which is:

$$\sigma^2 = \frac{1}{n-1} \sum_{i=1}^{n} (x_i - \mu)^2$$

In reality, for the normal distribution which we adopted for the measurements x_i, we must set:

$$\chi^2 = \frac{n\,s^2}{\sigma^2}$$

It has been shown that the variable χ^2 has the following distribution density (see [BOG 06]):

$$\varphi(\chi^2) = \frac{1}{2^{n/2}\Gamma(n/2)} x^{\left(\frac{n}{2}-1\right)} e^{-\frac{x}{2}}$$

This function can only be integrated numerically. Fortunately, there are published tables giving the value of the function $\varphi(x^2)$ ([BOG 06; p. 355] and also [SPI 74; p. 259]).

Figure 3.4 shows the shape of the variations of the function φ (x, n) for $n = 20$.

3.3.4. Acceptability criteria for the mixture [RAA 90]

1) Let us set:

$$\sigma_x^2 = s_x^2 \big|_{z \Rightarrow \infty}$$

This value σ_x^2 remains unknown. σ_x is the standard deviation.

We can only choose an upper bound σ_1^2 for σ_x^2 if $\sigma_1^2 \geq \sigma_x^2$.

2) However, if we choose $\sigma_1 < \sigma_x^2$, then we must have:

$$s_x^2 \leq s_{xo}^2$$

or indeed:

$$\chi^2 = z\frac{s_x^2}{\sigma_x^2} \leq z\frac{s_{xo}^2}{\sigma_1^2} = \chi_{o1}^2$$

We have replaced σ_x^2 with σ_1^2, because we do not know σ_x^2.

The inequality can be written thus:

$$s_x^2 \leq s_{xo}^2 \left(\frac{\sigma_x}{\sigma_1}\right)^2$$

The value of the term in parentheses is greater than 1, and we do not necessarily have:

$$s_x^2 \leq s_{xo}^2$$

but we can always write:

$$\sigma_x^2 = \sigma_2^2 > \sigma_1^2$$

Generally, therefore, if we refer to Figure 3.5, we shall have:

$$s_{x2}^2 > s_{x1}^2$$

The result of this is that the curve $\varphi\left(\frac{s_x}{\sigma_1}\right)_2$ will be shifted to the right in relation to the curve $\varphi\left(\frac{s_x}{\sigma_1}\right)_1$.

3) Let us examine the curve φ_1 for which $\sigma_1^2 = \sigma_x^2$. The vertical D gives the limit which must not be surpassed in order for $s_x \leq s_{xo}$. The surface α corresponds to the crossing of that limit, and the corresponding system will be rejected, but wrongly so, as $\sigma_x^2 = \sigma_1^2$. This is the first-species error.

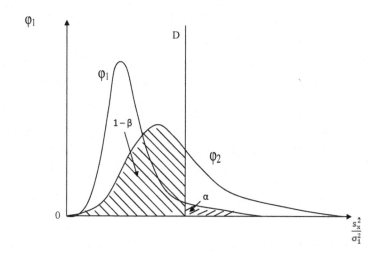

Figure 3.5. *Probability densities*

4) Now suppose that $\sigma_x^2 = \sigma_2^2 > \sigma_1^2$.

Because $\sigma_x^2 > \sigma_1^2$, we generally have $s_{x2} > s_{x1}$, and the curve φ_2 will be shifted to the right in relation to the curve φ_1.

The probability $(1 - \beta)$ corresponds to an acceptance of the system, but this is erroneous because σ_x^2 is not less than σ_1^2. This is the second-species error.

NOTE.–

If we shift the limiting line D, we see that the probabilities α and $(1 - \beta)$ vary in opposite directions.

If we increase the number of samples, the probability $(1 - \beta)$ decreases whilst the probability α remains constant.

NOTE.–

[LAN 72] describes original procedures for acceptance or rejection. These procedures are not particularly simple. They have been employed by the company Ciba-Geigy. Undoubtedly, this procedure is still in use today.

3.4. Evolution of the DS over the course of mixing

3.4.1. *Mechanisms of mixing*

The phenomenon of mixing in a DS, in all cases, requires the application of energy by stirring. Such stirring may give rise to the following phenomena:

– convection;

– shearing;

– diffusion.

Diffusion in a liquid is spontaneous, as long as there is a concentration gradient in place. This is different with fluidization, where the energy is provided by the gas, which disturbs the particles and gives rise to the motions of convection and diffusion.

Diffusion exists in horizontal rotary drums, where disturbance is caused by the dropping of the particles and their re-ascent on the wall of the drum. Here again, there are movements of convection and diffusion.

If we use a palette thruster, plunged into a dense-phase DS, we cause motions of convection, which create a certain degree of shearing between the domains moving at different velocities. Convection and shearing are therefore indissociable from one another [BRI 76].

Percolation is the assisted diffusion of fine particles between larger particles. This progression of fine particles between the large particles is generally due to the forces of mass (gravity). However, shearing is necessary for percolation, which is why percolation plays a part in the flow along an inclined plane.

Percolation, diffusion and shearing may cause *segregation*, whereas convection does not.

3.4.2. *Motions in a DS*

With a single-line simulation, [GYE 99] shows that the displacements imposed within a DS are only of two natures:

– diffusion;

– convection/shearing.

Diffusion can only take place if the particles are provided with energy from the external environment. It occurs on a newly-created surface either within the DS or on its surface.

In a liquid, the provision of energy is not necessary, because molecular agitation is spontaneous, and is determined by the temperature. [BRI 76] reproduces Einstein's calculation of the diffusivity of large molecules (or of small particles) in a liquid. This movement is known as Brownian motion.

Diffusive transport is effective if, after having simultaneously injected a given number of particles at the same place, the variance of the displacement is proportional to the elapsed time. However, this has never been tried in the case of a DS.

Convection is the transport of groups of numerous particles from one place to another. It may be caused by gravity or by a mechanical agitation.

Within a DS, shearing and convection are inseparable. There can be no shearing without convection, and convection very generally causes shearing.

Particularly if the product is cohesive, motion can only begin with ruptures occurring along shearing surfaces (faults) whose thickness is equal to around ten times the particle diameter.

The distance between two faults depends on the velocity of the shearing motion, the pressure of prior consolidation and the composition of the mixture of DS.

Let us examine two case studies.

For a *palette stirrer* immersed in a DS, the resistant torque increases with depth, up to a certain point. It increases proportionally to the height of the palettes, and much more quickly with their length. A kernel of DS rotates with the palettes and remains separated from the rest of the DS by a rupture surface.

A palette moving perpendicularly to itself at the surface of a DS encounters resistance which is proportional to its length. To extrapolate from one palette to another, we must preserve geometric similarity between d_p and the dimensions of the palette.

3.4.3. *Evolution of a mixture and variance*

Numerous definitions have been put forward in this area. If we base our reasoning on the variance, we can write the following for the quality M of a mixture, i.e. its homogeneity (which could alternatively be referred to as its mixity).

$$M = 1 - \frac{s}{s_0} \quad \text{or indeed} \quad M = 1 - \frac{s^2}{s_0^2}$$

As a function of the time t, it is commonly held that:

$$M = 1 - e^{-kt}$$

The measured variance s decreases regularly over the course of the operation of mixing of a load. However, it may arise that, towards the end of the process, segregation occurs, and M begins to decrease, thus passing through a maximum (see section 3.4.4).

In addition, Poole *et al.* [POO 64], rather than accepting an exponential decrease, divide the mixing operation into two stages:

– from 0 to 10 s: quick initial mixing;

– $t > 10$ s: the variance s decreases to the power -0.20 of the time.

We can adapt the definition of M to a continuous operation, and set:

$$s_0 = \quad \left\{ \begin{array}{l} \text{the initial variance for the processing of a load} \\ \text{or else} \\ \text{the variance at input for a continuous operation} \end{array} \right.$$

$$s_1 = \quad \left\{ \begin{array}{l} \text{the final variance for the processing of a load} \\ \text{or else} \\ \text{the variance at output for a continuous operation} \end{array} \right.$$

3.4.4. Evolution of variance according to [ROS 59]

Let us take the following expression for the degree of homogeneity:

$$M = 1 - \frac{s(t)}{s_0}$$

[ROS 59] proffers:

$$\frac{dM}{dt} = A(1-M) - B\varnothing$$

He shows that:

$$M = \eta \left[1 - \left(\pm \left(1 \mp \frac{B}{A} \right) e^{-At/2} + \frac{B}{A} \right)^2 \right]$$

The shape of M is given by Figure 3.6.

Depending on the value of the parameters of A and B, the equation M may or may not pass through a maximum, proving that, beyond a certain point, segregation may manifest itself.

However, Rose's calculations are not linked to the properties of the divided solids in question.

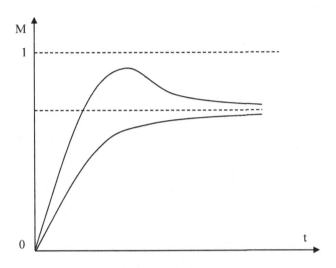

Figure 3.6. *Evolution of degrees of homogeneity*

3.4.5. *Cohesion and particle diameter*

Taking the mean of the indications given by [BRI 76] and by [CHO 79], we shall write:

$d_p > 150$ μm   Flows freely

25 μm $< d_p < 150$ μm   Low flowability. Cohesion manifests itself.

$d_p < 25$ μm   Zero flowability. Notable cohesion

In a gaseous atmosphere, cohesion is attributable, in particular, to van der Waals forces or to the humidity. The smaller the particles, the more predominant are the van der Waals forces in relation to the weight of the particles.

[PIE 67] give the geometry of the liquid menisci linking two or more particles. The forces of adhesion (of adherence) are calculated.

When, as is generally the case, the active principle is fine, it may adhere to the walls by van der Waals attraction. Palettes are then very helpful in unsticking what has remained attached to the wall. Otherwise, the mixture obtained would be poor in active ingredient.

3.4.6. *Analytical attempts to express a mixing operation*

The equations directly describe the evolution of the concentration of a marker.

1) Planck's equation [PLA 17]

The so-called Kolmogorov equation was established by Planck in 1917, as mentioned in Ruzbehl *et al.* [RUZ 82]. This equation is written:

$$\frac{\partial q(x,t)}{\partial t} = -\frac{\partial}{\partial x}\left[T(x,t)q(x,t)\right] + \frac{\partial^2}{\partial x^2}\left[D(x,t)q(x,t)\right]$$

q: mass fraction of the control product.

The Kolmogorov equation combines the action of convection by the parameter $T(x, t)$ and the action of diffusion by the parameter $D(x, t)$ (the letter T stands for "transport").

2) Case of a fixed load

Rotary tubes correspond to $T(x, t) = 0$ and to $D(x, t) =$ const. We then obtain Fick's second equation:

$$\frac{\partial q(x,t)}{\partial t} = D\frac{\partial^2 q(x,t)}{\partial x^2}$$

If the rotary tube is divided into two parts by a vertical plane perpendicular to its axis, and each part is filled with a different DS, the diffusion equation well expresses the evolution of the concentration q, on the basis of which we can calculate the evolution of the variance and compare it with the variance experienced. The authors' Figure 4 shows that they are identical.

3) Continuous operation of the rotary tube

As the regime is permanent, the equation becomes:

$$\frac{d}{dx}\left[T(x)q(x)\right] = \frac{d^2}{dx^2}\left[D(x)q(x)\right]$$

The solution to this equation is:

$$q(x) = K \exp \int \frac{T(x)}{D(x)} dx$$

The concentration profile of a tracer may be inverted if there is segregation:

– the concentration decreases from the inlet to the outlet;

– then that concentration increases from the inlet to the outlet.

4) Planck's equation (also known as the Kolmogorov equation) is written thus (see [FAN 79]):

$$\frac{\partial c}{\partial t} = \frac{\partial^2 (Dc)}{\partial x^2} - \frac{\partial (Vc)}{\partial x}$$

$c(x, t)$: local and instantaneous concentration of the marker

From this, the authors deduce their equations 29 and 30:

$$D_h(x_i) = \frac{1}{\Delta t_k} \sum_{j=1}^{n} \frac{1}{2} (x_j - x_i)^2 \frac{P_{ij}}{n}$$

$$V_k(x_i) = \frac{1}{\Delta t_k} \sum_{j=1}^{n} \frac{1}{2} (x_j - x_i) \frac{P_{ij}}{n}$$

We evaluate D_h and V_k for increasingly long values of Δt_k, and we then merely need to extrapolate to $\Delta t = 0$ to find $D(x_i)$ and $V(x_i)$.

The authors show experimentally that Planck's equation, with D = const. and with V dependent on x rather than on t, can be used to express the behavior of a continuous tubular mixer.

5) The stochastic matrix method

The mixer is divided into n elementary domains, and the particles in each domain disperse into the n-1 other domains in the wake of an elementary operation.

An elementary operation may be the complete emptying of a hopper into an identical hopper placed underneath the first. Then, the positions of the two hoppers are interchanged. Another elementary operation could be a revolution (a turn) of a drop/rise device.

The authors introduce a so-called stochastic matrix P_{ij}, each element of which distributes the concentration at the point i between the n points j following an elementary operation (j can be equal to i).

After n elementary operations:

$$C_n = P^n C_o$$

C: vertical vector of concentrations in the n places.

The quality of the mixture is defined here by:

$$M = 1 - \frac{s}{s_0}$$

$$s^2 = \frac{1}{w} \sum_{i=1}^{w} (x_i - \overline{x})^2 \qquad s_0 = \overline{x}(1 - \overline{x})$$

[BOS 85a] determine the elements P_{ij} of the matrix P by placing a tracer into just one domain: the domain i.

The same authors [BOS 85b] generalize the method to a number of components greater than 2.

3.4.7. Duration of the operation of mixing

We gain time and save energy if we can perform mixing quickly. However, the duration of the operation may be long if the reciprocal motion of the particles is hampered, and this may be due to two primary causes:

– the existence of aggregates;

– mutual friction between the particles.

1) Aggregates

The existence of aggregates may be attributable either to electrostatic forces or to excess humidity.

If certain particles are insulators (plastics) and if the operation of mixing is long enough, we see the appearance of electrostatic forces. A slight humidification (by 1–2% in terms of mass) will eliminate these forces and enable us to process the mixture by moderate stirring.

However, if the humidity is excessive (greater than 4%), the mutual cohesion of the particles is increased by capillary forces, and again we see the formation of aggregates. The only solution in this case is slight drying.

2) Mutual friction

The mutual friction between the particles is increased if the majority component is damp, which increases the interparticle contact surfaces (and can even engender stable aggregates). It is possible to combat the softening of the component by cooling down the device, but this may be costly.

The inter-particle friction may also be significant if the majority component exhibits a rough surface and/or has an angular form or one far removed from a sphere. We can then minimize friction by the adjunction of adjuvants, which might include:

– stearates which, in a small quantity, are veritable lubricants but, in too great a proportion, cement the grains to one another, causing the formation of aggregates;

– talc or fine silica; the PSD of these products is less than 10μm and, added in a small quantity, they affix to the particles being mixed, by van der Waals forces, and also act as a lubricant.

The best way, though, of minimizing inter-particle friction is to prevent contact between the particles. Pneumatic methods are therefore all recommended, provided the limiting velocity of the drop of the particles in the mixture through the air remains in a sufficiently narrow range.

3) Practical data

Discontinuous operations by pouring, dropping or stirring, carried out on one or a few cubic meters of mixture, last for around 15 minutes. On the other hand, if we wish to process hundreds of cubic meters (which is only possible by pneumatic techniques), the duration of the operation will vary between half an hour and four hours. Indeed, whilst it is relatively easy to obtain a homogeneous mixture in a small device, the same is not true in a large-capacity mixer, as the variations in composition are greater because the distance between the elementary domains is significantly larger.

Very approximately, according to Poole *et al.* [POO 64], when the weight fraction of the dilute DS decreases from 1/10 to 1/1000, the time needed to obtain random mixity decreases by two orders of magnitude.

We may legitimately wonder whether, at the same level of energy expenditure, we need to operate quickly with high power or slowly with a low level of power. Experience shows that, at least in drop/pour or moderate stirring systems, it is preferable to act quickly, because this limits attrition and segregation by percolation.

3.5. Mixers (practical data): choice of device

3.5.1. *The operation of homogenization (mixing)*

The purpose of the operation of mixing is to make the position of the different particles within the mixture as random as possible, regardless of their nature. Therefore, we have to modify their original positions as, to begin with, they are ordered.

However, a divided solid does not flow in the same way as does a liquid and, to facilitate relative motion of the particles, we need to increase the volume of empty spaces between them, working against the effect of gravity. At least four methods are used for this purpose:

1) dropping (or pouring), alternating with rise/ascent;

2) convection/shearing (moderate internal stirring);

3) violent convection, i.e. milling.

3.5.2. *Devices employing dropping (or pouring) and raising*

Such devices operate discontinuously.

The load is poured in the form of a cascade or of a shower. Neighboring jets expand, interpenetrate with one another and mix. This action of mixing is fairly gentle, and the energy consumed is moderate – between 2.5 and 4 kW per m^3 apparent of solid. This method is only applicable for products which do not tend to segregate under the influence of gravity.

The devices used are chambers of various forms, rotating around a horizontal axis which traverses them. Certain devices have the shape of a double cone or a fork (i.e. the shape of a Λ or an upside-down V). The mixture is brought up by friction on the ascending wall, and falls back on itself, usually in a cascade. An optimum value of the filling rate needs to be preserved. Indeed, if that rate is too low, the mass slides in a block against the wall and, if it is too high, the product behaves as it does in a fixed bed, and also rotates in a block.

Cleansing from one operation to the next is easy, and the walls experience negligible erosion by the divided solid. The rotation frequency must remain less than 0.6 times the critical velocity of centrifugation.

$$N = 0.6 N_C$$

[WAN 74] propose extrapolation methods for drop/pour and rise devices.

3.5.3. *Mixing in a screw transporter*

The mixing performed by a screw is slight.

[TSA 94] offer advice on how to increase the effectiveness of a screw in terms of the operation of mixing – particularly by exploiting the shape of the different parts of the device.

The mixing of different DSs is more effective if the particles of the two products have the same properties.

The effectiveness of the mixing increases with the speed of rotation of the screw, and all the more so if the screw is replaced with inclined palettes.

3.5.4. *Rotary cylinder*

[MU 80] put forward a model for the distribution of the residence time of the product in the cylinder. They determine a rate of bypass of the device by the DS and a recirculation rate.

The authors determine the number of equivalent stages to evaluate the degree to which the cylinder differs from a simple mixer.

[DAN 53a], in his equation 46, gives the variance of a composition as a function of the distribution of length of stay and the autocorrelation function. This result is undoubtedly applicable to the rotary cylinder.

[MUL 67] studied the concentration profile in a rotary tube in continuous operation.

3.5.5. *Devices with moderate internal stirring (convection/shearing)*

These devices may run continuously or process a batch at a time. Inside the chamber containing the mixture is a rotary device (a rotor) which stirs the solid mass. Depending on the design of the rotor and its rotation speed, the power consumed ranges from 7 to 25 kW per m^3 apparent of mixture. These devices are particularly apt for products which exhibit a tendency toward segregation under the influence of gravity.

We distinguish between:

1) devices stirred internally by plowshare blades [SCH 80];

2) the planetary screw cone: this is a load-based device composed of a vat in the shape of a cone, whose point is at the bottom. A screw rotates on itself, sweeping the generators of the cone. This screw brings the product upwards, before it falls back down under the influence of gravity, which creates internal recirculation. This type of mixer is well suited for flowable products;

3) the ribbon mixer: this device can be used continuously or on a load-by-load basis. It contains two helical ribbons with the same external diameter, rotating around a shared horizontal axis. They are designed to cause the product to circulate in opposite directions. In continuous-flow devices, the ribbons are of different widths: the broad ribbon transports the product and the thin ribbon shears it. The ribbon mixer is useful for fibrous or sticky products. If we are dealing with a pourable mixture, the ribbons must be broad enough to entrain the solid and have a sufficiently powerful stirring action.

3.5.6. *Violent internal stirring devices (mills)*

Here we are dealing with certain continuous-flow mills:

– the pin crusher ensures both the extensive particle size reduction and the thorough mixing of the components fed in. It is possible to sweep the inside of the machine with a stream of air to evacuate the fine particles;

– the running mill stone mill has also been used for mixing operations, on condition that it be equipped with scraping knives.

Mills are recommended if the aim is to destroy the clumps of material. Indeed, the van der Waals forces are stronger when the particles are finer. Thus, below 50 µm, particles of the same nature may aggregate and clump together, which renders mixing difficult by a drop-based or pouring method or by moderate stirring.

NOTE (incorporation of a liquid into a solid).–

This operation is used, in particular, to humidify a product which is liable to release dust during later handling. The mixer with two horizontal parallel screws turning in the same direction is very apt for this purpose, i.e. for sticky mixtures. Indeed, the rotation of the two screws in the same direction causes a shearing effect and prevents the product, having become sticky because of the humidification, rotating with the screws without progressing. The consumption is around 7 kW/m^3.

In a pharmacy, the bladed basin is used: this is a cylindrical container whose vertical axis has a shaft with scrapers in the form of plowshares, which actively stir the mixture. This system consumes around 25 kW/m^3.

3.5.7. *Energy consumed*

According to Ruzbehl *et al.* [RUZ 82], the power consumed by a mixer (in kW) increases exponentially with the density of the product (see authors' Figure 5).

The work (in joules) decreases with the size of the particles (author's Figure 7). This is the work put to use in a mixer.

The work expended in a mixer passes through a minimum as a function of the ratios:

$$\frac{\text{diameter of particles of an active principle}}{\text{diameter of particles of the excipient}}$$

and

$$\frac{\text{density of the active}}{\text{density of the excipient}}$$

[SCH 80] studied the resistant torque of the full mixer.

3.5.8. *Choice of mixers*

Numerous of mixers are described by [MUL 67, RIE 79] and [MÜL 81].

We distinguish:

1) Devices including a gravity-based drop followed by re-ascent

In double-cone mixers or rotary-cylinder mixers, the solid climbs up because it is entrained by the wall.

In Λ-shaped devices, the DS falls into one of the branches of the Λ, which is then emptied freely as the device rotates.

In the planetary screw system, the screw brings the DS upwards, and the solid then falls back down under the influence of gravity.

Drop/rise mixers cause the diffusion of the DS at the end of its fall.

2) Convection/shearing devices

These include ribbon mixers, and horizontal cylinders with palettes which rotate around the axis of the cylinder. Vertical-axis mixers with palettes are rarely found, because they expend a great deal of energy, as the palettes are immersed in the DS. It is judicious to replace the palettes with plowshare-shaped objects.

Convection devices inevitably cause shearing. It is for this reason that they are sometimes known as convection/shearing devices.

3) *From the standpoint of effectiveness*, according to [CHO 79]:

– if the excipient is non-cohesive, drop/rise devices are effective;

– if the excipient is, itself, cohesive, a convection-shearing device is preferable;

– if a component has a tendency to form aggregates, a convection/ shearing device must be used.

NOTE.–

It is sometimes preferable to make the mixture *ab initio* – i.e. to make it directly in the correct proportions by adjusting the feed flowrates and putting inlets in place so that the jets of DS interpenetrate – rather than, *a posteriori*, attempting to mix completely separated DSs in a mixer.

3.6. Segregation

3.6.1. *Principle*

Consider a binary mixture, i.e. a mixture composed of two divided solids of different natures. If segregation is complete, the two DSs are separated, situated in two different parts of the device in which the divided solids are being treated.

Generally, though, segregation is not total – merely partial – which can seriously skew the results of a mixing operation.

There are a variety of factors which can cause segregation, which we shall now examine. In the coming discussion, we have to consider two different sizes of particles: *we distinguish between large particles and fine particles*. Indeed, the dimension of the particles and gravity are the most important parameters in terms of segregation.

3.6.2. *Influence of a vertical vibration*

In a plane where the coordinates are the frequency of vibration and the height in relation to the base of the recipient containing a DS, [WIL 63] studied the motion of a large particle whose diameter is at least ten times that of the particles of the DS. The plane is divided into three domains:

– rest;

– above;

– below.

On the basis of this representation, the author describes the possible motion of the large particles.

3.6.3. *Spilling (pouring) of a loose DS onto a pile*

[WIL 63], in this operation, studied the concentration of fine particles of the mixture at the periphery (at the base) of the pile.

Initial composition	Periphery
50%	1%
80%	20%

Indeed, the edge of the pile is greatly enriched with large particles.

Indeed, when we pour a pourable divided solid, the horizontal distance traveled by the particles is an increasing function of their diameter, as the horizontal deceleration due to air resistance is greater for finer particles.

3.6.4. *Dropping under the influence of gravity*

Bridgwater *et al.* [BRI 69] dropped fine particles which traverse a bed of large particles. The laws of radial diffusion apply at the end, and the diffusion equation, integrated by [WIL 04], can be used to calculate the dispersivity of the fine particles. This dispersivity is a linear decreasing function of the coefficient of restitution of the fine particles, and is independent of the diameter of those fine particles, provided that diameter remains less than a certain limit.

[WAN 77] apply the model of a Markov chain to the drop of a mixture containing particles of a tracer. The quality of the mixture is:

$$M = 1 - \frac{s^2}{s_o^2}$$

s was the variance of the mixture obtained at the base of the device;

s_o was the variance of the mixture when fed in.

The drop took place in a helical static mixer. The calculations do indeed reflect reality.

3.6.5. *Gravity-based flow on an inclined plane*

[SHI 79, SHI 82, SHI 84, SHI 85] divides the layer of DS into three parts, parallel to the inclined plane:

– the fixed, lower layer, said to be *enriched* because the DS is a mixture of large and fine particles, and the latter percolate between the former;

– the intermediary layer, said to be *segregating*, which is traversed by the fine particles from the upper layer. The intermediary layer is subject to a velocity gradient parallel to the inclined plane;

– the mobile upper layer, *impoverished* of its fine particles.

The author calls this the "layering model".

The mobile layer is divided into mobile blocks by vertical sections. The common index of the blocks is m.

The fixed inclined plane is divided into fixed zones whose common index is n.

Figures 11 to 17 in [SHI 82] show that the calculations accurately reflect reality.

3.6.6. *Effect of convection/shearing*

Bridgwater *et al.* [BRI 78] showed that the percolation of fine particles is controlled by shearing, by gravity and by the ratio of the diameter of fine particles divided by that of the large particles.

[COO 79] showed that the following relation exists:

$$Ln\left(\frac{u}{\dot{\gamma}d_g}\right) = \text{linear decreasing function of}\left(\frac{d_f}{d_g}\right)$$

u: vertical velocity of percolation of the fine particles: $m.s^{-1}$;

$\dot{\gamma}$: velocity of shear displacement: s^{-1};

d_f: diameter of fine particles: m;

d_g: diameter of large particles: m.

Bridgwater *et al.* [BRI 85] use the fact that the majority of a DS subject to convection/shearing moves in the form of coherent blocks, separated by regions of significant displacement (which we shall call rifts). They examine the case where the rifts is vertical and where the shearing displacement is horizontal and perpendicular to the rifts.

The following notation is used:

v: horizontal velocity: $m.s^{-1}$;

t: time: s;

d_g: diameter of the large particles: m.

The authors set the fundamental equation:

$$p = \frac{dz}{dt} = -p^* d_g \frac{dv}{dz}$$

[3.5]

z: altitude: m;

p*: dimensionless constant;

p: percolation rate: $m.s^{-1}$.

The authors accept that:

$$v = v_o th\left(\frac{z}{z_o}\right)$$

v_0: horizontal velocity of the upper part of the fault: $m.s^{-1}$.

The velocities v and v_0 are perpendicular to the rifts. This is made possible by the use of a circular shearing cell.

The authors integrate fundamental equation [3.5] and note that the result of their calculations is confirmed by experience (see their Figures 7 and 9). These results are expressed by the function c(z), where c is the volumetric fraction of fine particles.

Bridgwater *et al.* [BRI 85] examine, in detail, the concentration of fine particles in the rifts (their Figure 9).

NOTE.–

The large particles move toward those regions where the shear displacement is greatest [FOO 83]. The migration rate is proportional to the gradient of the shearing rate.

3.6.7. *Mathematical synthesis [CAR 81]*

Cardew describes percolation, starting with:

– either an equation involving the probability of passage from one horizontal level to another in the DS subject to shearing;

– or else Planck's equation [PLA 17], which is also known as Kolmogorov's equation or Levenspiel and Bischoff's equation, and is written:

$$\frac{\partial c}{\partial t} + v\frac{\partial c}{\partial z} = D\frac{\partial^2 c}{\partial z^2}$$

The value found for the Péclet number closely matches the experience of reality. The author looks at the following case studies:

– descent of the fine particles through the coarse ones (percolation under the influence of gravity);

– random motion of autodiffusion between particles of the same sizes.

3.6.8. *Ways of combatting segregation*

A moderate relative humidity of the environment favors the adsorption of water to the surface of the particles, which heightens the intensity of the forces of mutual cohesion. Beyond a critical value of around 65–80%, we see the appearance of capillary bridges between the particles, which can cause the forces of cohesion to increase tenfold. The mixture then takes longer to obtain, but it is more stable, and does not give rise to percolation. In practical terms, we spray the solid with 1–2% mass water, which can decrease the mixture's characteristic standard deviation by a factor of around 7.

Processing the mixture in a mill is another way to remedy segregation by percolation, when that segregation is essentially due to a ratio d_f/d_g that is much lower than 1.

Finally, to limit percolation in an operation with moderate stirring, it may be useful to place the percolating component (fine, dense), at the top of the load to begin with, and limit the mixing time:

– if the fine particles are much smaller in size than the coarser grains – i.e. if d_f/d_g is much less than 1 – then the larger grains are simply coated by the finer particles. The latter are held in place by van der Waals forces;

– if the particles have a rough surface, this hampers relative slippage, and it will take longer to obtain the mixture but, once finished, it will be more stable.

3.6.9. *Conclusions on segregation*

For numerous systems, we see that it is impossible to obtain a perfectly random mixture. The reason for this is the existence of segregation. Rose's kinetic model seeks to take account of this fact, but is incapable of determining which mechanism of homogenization is at work.

Segregation can often take place by the migration of fine particles through the spaces left between the larger particles. "Fine particles move amongst the grains". This displacement is known as percolation (from the Latin *percolare*, which means to filter).

Percolation is facilitated by sphericity of the particles, because a sphere has a high level of mobility. Angulosity hampers segregation, and flattened or elongated forms hinder mobility and therefore segregation.

Percolation decreases if the shearing is fast, because percolation is a rather slow phenomenon and does not have enough time to take place. However, it can continue once the DS has reached its resting state.

[GYE 99] states that shearing and diffusion can cause segregation at the same time as homogenization. On the other hand, convection acts against segregation.

According to [RUZ 82], stabilizing additives can inhibit the movement of the particles. Depending on the case, either water or oil will need to be added immediately before maximum homogeneity is reached. As soon as this happens, mixing must be halted.

APPENDICES

Appendix 1

Mohs Scale

Nature of the divided solid	Mohs hardness
Wax	0.02
Graphite	0.5–1
Talc	1
Diatomaceous earth	1–1.5
Asphalt	1.5
Lead	1.5
Gypsum	2
Human nail	2
Organic crystals	2
Flaked sodium carbonate	2
Sulfur	2
Salt	2
Tin	2
Zinc	2
Anthracite	2.2
Slaked lime	2–3
Silver	2.5
Borax	2.5
Kaolin	2.5
Litharge (yellow lead)	2.5
Sodium bicarbonate	2.5
Copper (coins)	2.5

Slaked lime	2–3
Aluminum	2–3
Quicklime	2–4
Calcite	3
Bauxite	3
Mica	3
Plastic materials	3
Barite	3.3
Brass	3–4
Limestone	3–4
Dolomite	3.5–4
Siderite	3.5–4
Sphalerite	3.5–4
Chalcopyrite	3.5–4
Fluorite	4
Pyrrhotite	4
Iron	4–5
Zinc oxide	4.5
Glass	4.5–6.5
Apatite	5
Carbon black	5
Asbestos	5
Steel	5–8.5
Chromite	5.5
Magnetite	6
Orthoclase	6
Clinker	6
Iron oxide	6
Feldspar	6
Pumice stone	6
Magnesia (MgO)	5–6.5
Pyrite	6.5
Titanium oxide	6.5
Quartz	7

Sand	7
Zirconia	7
Beryl	7
Topaz	8
Emery	7–9
Garnet	8.2
Sapphire	9
Corrundum	9
Tungsten carbide	9.2
Alumina	9.25
Tantalum carbide	9.3
Titanium carbide	9.4
Silicon carbide	9.4
Boron carbide	9.5
Diamond	10

Thus, materials can be classified according to their hardness:

Soft   1 to 3

Fairly soft   4 to 6

H   7 to 10

Appendix 2

Apparent Density of Loose Divided Solids (kg.m^{-3})

I – Plant products

Nature of product	Grains or seeds		Flour	
Flax	720		430	
Maize	720		640	
Cotton	530		400	
Soya	700	(shredded)	540	
Coffee	670	(green)	450	(roasted, ground)
Wheat	790			
Barley	620			
Rye	720			
Rice	800			
Oats	410			
Clove	770			

II – Inorganic natural products

Nature of product	Grainy		Powders (fine milling)	
Bauxite	1360	(mill run)	1090	
Gypsum	1270		900	
Kaolin	1024	(crushed)	350	(<10μm)
Lead silicate	3700		2950	
Quicklime	850		430	
Limestone	1570		1360	
Phosphate	960		800	

Wood remnants	350	(shavings)	320	(sawdust)
Sulfur	1220		800	
Iron	4950	(bearings)	2370	(filings)
Slate chippings	1390		1310	
Sodium carbonate	1060		480	
Coke	490		430	

III – Manufactured products

Powdered sodium bicarbonate	690
Borax	1700
Catalyst (fluid catalytic cracking)	510
Ash	700
Wood charcoal (grains)	420
Coal (mill run)	900
Coal (classified)	700–800
Cement (clinker)	1400
Cement (Portland)	1520
Chips of shredded copra	510
Copra chips output from a pressure screw	465
Powdered dolomite	730
Soap scales	160
Milled feldspar	1600
Gravel	1500
Slag	2000
Ground mica	210
Ground bone	1200
Phtalic anhydride shavings	670
Glass pearls	1400
Iron oxide pigment	400
Zinc oxide pigment	320
Buckshot	6560
Potatoes	700
Rubber shavings	370
Sand	1350–1500
Salt	1200
Crystallized sugar	830
Crystallized copper sulfate	1200
Powdered superphosphate	810

Bibliography

[ALL 00] ALLEN H.S., "The motion of a sphere in a viscous fluid", *Philosophical Magazine*, vol. 50, pp. 323–338, 1900.

[ANO 83] ANONYMOUS, "Mélange et agitation", *Information Chimie*, no. 236, pp. 139–149, 1983.

[ARB 64] ARBITER N., HARRIS C.C., STEININGER J., "Power requirements in multi-phase mixing", *Transactions of Society of Mining Engineers*, vol. 229, pp. 70–78, 1964.

[BER 72] BERKEVITCH I., "New ideas in mixing pharmaceuticals and other fine powders", *Manufacturing and Aerosols News*, vol. 43, no. 4, pp. 36–37, 1972.

[BOG 06] BOGAERT B., *Probabilités pour scientifiques et ingénieurs*, De Boeck, 2006.

[BOS 85a] BOSS J., DABROWSKA D., "Stochastic model of mixing during discharging of granular materials from a bin. I: two-component system", *Journal of Powder and Bulk Solids Technology*, vol. 9, no. 4, pp. 1–11, 1985.

[BOS 85b] BOSS J., DABROWSKA D., "Stochastic model of mixing during discharging of granular materials from a bin. II: multicomponent system", *Journal of Powder and Bulk Solids Technology*, vol. 9, no. 4, pp. 12–14, 1985.

[BOU 67] BOURNE J.R., "Variance-sample size relationships for incomplete mixtures", *Chemical Engineering Science*, vol. 22, pp. 693–700, 1967.

[BOW 68] BOWEN R.L., "Agitation intensity: key to scaling up flow-sensitive liquid systems", *Chemical Engineering*, pp. 159–168, 1985.

[BRI 69] BRIDGWATER J., SHARPE N.W., STOCKER D.C., "Particle mixing by percolation", *Transactions of the Institution of Chemical Engineers*, vol. 47, pp. T114–T119, 1969.

[BRI 76] BRIDGEWATER J., "Fundamental powder mixing mechanisms", *Powder Technology*, vol. 15, pp. 215–236, 1976.

[BRI 78] BRIDGWATER J., COOKE M.H., SCOTT A.M., "Inter-particle percolation: equipment development and mean percolation velocities", *Transactions of the Institution of Chemical Engineers*, vol. 56, no. 3, pp. 157–167, 1978.

[BRI 85] BRIDGWATER J., FOO W.S., STEPHENS D.J., "Particle mixing and segregation in failure zones. Theory and experiment", *Powder Technology*, vol. 41, pp. 147–158, 1985.

[BRU 68] BRUN E.A., MARTINOT-LAGARDE A., MATHIEU J., *Mécanique des fluides*, Dunod, 1968.

[CAL 58] CALDERBANK P.H., "Physical rate processes in industrial fermentation – Part I. The interfacial area in gas–liquid contacting with mechanical agitation", *Transactions of the Institution of Chemical Engineers*, vol. 36, p. 443, 1958.

[CAR 65a] CARR R.L., "Classifying flow properties of solids", *Chemical Engineering*, vol. 72, p. 169, 1965.

[CAR 65b] CARR R.L., "Evaluating flow properties of solids", *Chemical Engineering*, vol. 72, p. 163, 1965.

[CAR 81] CARDEW P.T., "Percolation and mixing in failure zone", *Powder Technology*, vol. 28, pp. 119–128, 1981.

[CHO 79] CHOWHAN Z.T., LINN E.E., "Mixing of pharmaceutical solids I. Effect of particle size on mixing in cylindrical shear and V-shaped tumbling mixers", *Powder Technology*, vol. 24, pp. 237–244, 1979.

[COO 79] COOKE M.H., BRIDGWATER J., "Interparticle percolation: a statistical mechanical interpretation", *Industrial and Engineering Chemistry Fundamentals*, vol. 18, no. 1, pp. 25–27, 1979.

[DAN 52] DANCKWERTS P.V., "The definition and measurement of some characteristics of mixtures", *Applied Scientific Research. Section A, Mechanics, Heat*, vol. A3, pp. 279–296, 1952.

[DAN 53] DANCKWERTS P.V., "Continuous flow systems. Distribution of residence times", *Chemical Engineering Science*, vol. 2, no. 1, pp. 1–13, 1953.

[DUP 69] DUPRÉ A., *Théorie Mécanique de la chaleur*, Éditions Gauthier–Villars, Paris, 1869.

[DUR 99] DUROUDIER J.P., *Pratique de la filtration*, Hermès, 1999.

[DUR 16] DUROUDIER J.-P., *Liquid–Solid Separators*, ISTE Press, London and Elsevier, Oxford, 2016.

[EGE 85] EGERMANN H., KEMPTNER I., PICHLER E., "Effects of interparticulate interactions on mixing homogeneity", *Drug Development and Industrial Pharmacy*, vol. 11, pp. 663–676, 1985.

[EGE 89] EGERMANN H., "Ordered powder mixtures: reality or fiction?", *Journal of Pharmacy and Pharmacology*, vol. 41, pp. 141–142, 1989.

[EGE 91] EGERMANN H., ORR N.A., "Comments on the paper 'Recent developments in solids mixing' by L.T. Fan et al.", *Powder Technology*, vol. 68, pp. 195–196, 1991.

[EIC 97] EICHLER P., DAU G., EBERT F., "Geometry and mixing performance of gravity discharge silo mixers", *European Congress on Chemical Engineering*, Florence, Italy, 2, pp. 983–987, 1997.

[FAN 79] FAN L.T., SHIN S.H., "Stochastic diffusion model of non-ideal mixing in a horizontal drum mixer", *Chemical Engineering Science*, vol. 34, pp. 811–820, 1979.

[FAN 90] FAN L.T., CHEN Y.-M., LAI F.S., "Recent developments in solids mixing", *Powder Technology*, vol. 61, pp. 255–287, 1990.

[FOO 83] FOO W.S., BRIDGWATER J., "Short communication. Particle migration", *Powder Technology*, vol. 36, pp. 271–273, 1983.

[GER 81] GERICKE H., "Dosieren und Mischen von Schüttgütern im Chargen und kontinuierlichen Betrieb", *Aufbereitungs Technik*, vol. 22, p. 15, 1981.

[GYE 99] GYENIS J., "Assessment of mixing mechanisms on the basis of concentration pattern", *Chemical Engineering and Processing*, vol. 38, nos. 4–6, pp. 665–674, 1999.

[HAN 80] HANSFORD D.T., GRANT D.J.W., NEWTON J.M., "The influence of processing variables on the wetting properties of a hydrophobic powder", *Powder Technology*, vol. 26, pp. 119–126, 1980.

[HEL 75] HELDMAN D.R., *Food Process Engineering*, The Avi Publishing Company Inc., 1975.

[HID 95a] HIDAKA J., "Flowrate and fluctuations of wall pressure during discharge of granular materials from bin-hopper systems", *Kagaku Kogaku Ronbunshu*, vol. 21, no. 3, pp. 581–587, 1995.

[HID 95b] HIDAKA J., KANO J., SHIMOSAKA A., "Flow mechanism of granular materials discharging from bin-hopper systems", *Kagaku Kogaku Ronbunshu*, vol. 20, no. 3, pp. 397–404, 1995.

[JOH 70] JOHANSON J.R., "Solids material handling in-bin blending", *Chemical Engineering Progress*, vol. 66, no. 6, pp. 50–55, 1970.

[JOH 72] JOHNSON M.C.R., "Particle size distribution of the active ingredient for solid dosage forms of low dosage", *Pharmaceutica Acta Helvetiae*, vol. 47, pp. 546–559, 1972.

[KÄP 70] KÄPPEL M., SEIBRING H., "Leitungsbedarf und Mischzeit beim Rühren hochviskoser Flüssigkeiten mit dem Wendelrührer", *Verfahrenstechnik*, vol. 4, no. 10, pp. 470–475, 1970.

[KHA 43] KHANG S.J., LEVENSPIEL O., "The mixing-rate number for agitation-stirred tanks", *Chemical Engineering*, pp. 141–143, 1976.

[KOS 65] KOSSEN N.W.F., ABD HEERTJES P.M., "The determination of the contact angle for systems with a powder", *Chemical Engineering Science*, vol. 20, p. 593, 1965.

[LAC 43] LACEY P.M.C., "The mixing of solid particles", *Transactions of Chemical Engineers*, vol. 21, pp. 53–59, 1943.

[LAC 54] LACEY P.M.C., "Developments in the theory of particle mixing", *Journal of Applied Chemistry*, vol. 4, pp. 257–268, 1954.

[LAC 76a] LACEY P.M.C., MIRZA F.S.M.A., "A study of the structure of imperfect mixtures of particles. Part I. Experimental technique", *Powder Technology*, vol. 14, pp. 17–24, 1976.

[LAC 76b] LACEY P.M.C., MIRZA F.S.M.A., "A study of the structure of imperfect mixtures of particles. Part II. Correlational analysis", *Powder Technology*, vol. 14, pp. 25–33, 1976.

[LAN 72] LANGENBUCHER F., "Statistical analysis of the USP XVIII content uniformity sampling plan for tablets", *Pharmaceutica Acta Helvetiae*, vol. 47, nos. 2–3, pp. 142–152, 1972.

[LER 76] LERK C.F., SCHOONEN A.J.M., FELL J.T., "Contact angles and wetting of pharmaceutical powders", *Journal of Pharmaceutical Sciences*, vol. 65, p. 843, 1976.

[LER 77] LERK C.F., LAGAS M., BOELSTRA J.P. *et al.*, "Contact angles of pharmaceutical powders", *Journal of Pharmaceutical Sciences*, vol. 66, p. 148, 1977.

[MER 96] MERSMANN A., EINENKEL W.-D., KÄPPEL M., "Auslegung und Maßstabs-Vergrößerung von Rührapparaten", *Chemie Ingenieur Technik*, no. 23, pp. 953–996, 1975.

[MIL 44] MILLER S.A., MANN C.A., "Agitation of two-phase systems of immiscible liquid", *Transactions of American Institute of Chemcial Engineers*, vol. 40, pp. 709–745, 1944.

[MOH 57] MOHR W.D., SAXTON R.L., JEPSON C.H., "Mixing in laminar-flow systems", *Industrial and Engineering Chemistry*, vol. 49, no. 11, pp. 1855–1856, 1957.

[MU 80] MU J., PERLMUTTER D.D., "The mixing of granular solids in a rotary cylinder", *AIChE Journal*, vol. 26, no. 6, pp. 928–934, 1980.

[MÜL 67] MÜLLER W., RUMPF H., "Das Mischen von Pulvern in Mischern mit axialer Mischbewegung", *Chemie Ingenieur Technik*, vol. 39, nos. 5–6, pp. 365–373, 1967.

[MÜL 81] MÜLLER W., "Methoden und derzeitiger Kenntnisstand für Auslegungen beim Mischen von Feststoffen", *Chemie Ingenieur Technik*, vol. 53, no. 11, pp. 831–844, 1981.

[NAI 82] NAIMER N.S., CHIBA T., NIENOW A.W., "Parameter for a solids mixing/segregation model for gas fluidized beds", *Chemical Engineering Science*, vol. 37, no. 7, pp. 1047–1057, 1982.

[NEM 04] NEMENYI M., KOVACS A.J., "New simulation Technique for particle mixing of stored bulk materials", *A S A E/C S A E Annual International Meeting*, Ottawa, Ontario, Canada, 1–4 August 2004.

[OLD 83] OLDSHUE J.Y., *Fluid Mixing Technology*, McGraw Hill, 1983.

[PIE 67] PIETSCH W., RUMPF H., "Haftkraft Kapillardruck Flussigkeits-volumen und Grenzwinkel einer Flussigkeis brücke wischen zwei Kugeln", *Zeitschrift für Technische Chemie, Verfahrenstechnik und Apparatwesen*, no. 15, pp. 885–936, 1967.

[PLA 17] PLANCK M., "Ueber einen Satz der statistischen Dynamik und seine Erweiterung in der Quanten Theorie", *Preusische Akademie Wissenschaft Sitzung der Physik, Mathematik Klasse*, vol. 24, pp. 324–341, 1917.

[POO 64a] POOLE K.R., "Mixing powders to fine-scale homogeneity: studies of batch mixing", *Transactions of the Institution of Chemical Engineers*, vol. 42, pp. 305–315, 1964.

[POO 64b] POOLE K.R., TAYLOR R.F., WALL G.P., "Mixing powders to fine-scale homogeneity: studies of batch mixing", *Transactions of the Institution of Chemical Engineers*, vol. 42, pp. T305–T315, 1964.

[RAA 90] RAASCH J., SOMMER K., "Anwendung von Statistischen Prüfverfahren im Bereich der Mischtechnik", *Chemie Ingenieur Technik*, vol. 62, no. 1, pp. 17–22, 1990.

[RIE 79] RIES H.B., "Mischtechnik und Mischgeräte 2. Teil", *Aufbereitrings Technik*, vol. 20, no. 2, pp. 78–98, 1979.

[ROS 59] ROSE H.E., "Suggested equations relating to the mixing of powders and its application to the study of the performance of certain types of machine", *Transactions of the Institution of Chemical Engineers*, vol. 37, pp. 47–64, 1959.

[RUZ 82] RUZBEHL M., ALT C., LÜCKE R. "Untersuchungen über den Einfluss von Schüttgutgrös-sen in Mischprozess", *Aufbereitungs Technik*, vol. 23, p. 316, 1982.

[SCH 68] SCHWARTZBERG H.G., TREYBAL R.E., "Fluid and particle motion in turbulent stirred tanks", *IEC Fund*, vol. 7, pp. 6–12, 1968.

[SCH 80] SCHEUBER G., ALT C., LÜCKE R. "Untersuchung des Mischungsverlaufs in Feststoffmis-chern unterschidlicher Grösse", *Aufbereitungs Technik*, vol. 21, pp. 57–68, 1980.

[SCO 75] SCOTT A.M., BRIDGWAGER J., "Interparticle percolation: a fundamental solids mixing mechanism", *Industrial and Engineering Chemistry Fundamentals*, vol. 14, no. 1, pp. 22–27, 1975.

[SHI 70] SHINOHARA K., SHOJI K., TANAKA T., "Mechanism of segregation and blending of particles flying out of mass-flow hoppers", *Industrial and Engineering Chemistry Process Design and Development*, vol. 9, no. 2, pp. 174–179, 1970.

[SHI 72] SHINOHARA K., SHOJI K., TANAKA T., "Mechanism of size segregation of particles in filling a hopper", *Industrial and Engineering Chemistry Process Design Development*, vol. 11, no. 3, pp. 369–376, 1972.

[SHI 79] SHINOHARA K., "Mechanism of segregation of differently shaped particles in filling containers", *Industrial & Engineering Chemistry Process Design and Development*, vol. 18, no. 2, pp. 223–227, 1979.

[SHI 82] SHINOHARA K., "General analysis of particle-segregation mechanism in filling vessels by a screening layer model", *Journal of the Society of Powder Technology, Japan*, vol. 19, pp. 462–469, 1982.

[SHI 84] SHINOHARA K., MIYATA S.-I., "Mechanism of density segregation of particles in filling vessels", *Industrial and Engineering Chemistry Process, Design and Development*, vol. 23, pp. 423–428, 1984.

[SHI 85] SHINOHARA K., "Modelle zur Schüttgut-Entmischung bei der Befullung von Bunkern", *Aufbereitungs Technik*, vol. 26, no. 3, pp. 116–122, 1985.

[SOM 73] SOMMER K., "Messverfahren zur Bestimmung der Mischgüte von Schokolade und zur Beschreibung des Mischverlaufes in Knetern und Conchern", *Revue Internationale du Chocolat*, vol. 28, pp. 2–11, 1973.

[SOM 74] SOMMER K., "Varianz der stochastischen Homogeneität bei Körner-mischungen und Suspensionen und praktische Ermittlung der Mischgüte", *VDI Berichte*, vol. 46, no. 218, pp. 415–428, 1974.

[SOM 75] SOMMER K., "Das optimal Mischgütemass für die Praxis", *Chemie Technik*, vol. 4, pp. 347–350, 1975.

[SOM 76] SOMMER K., "Mischgüte pulverförmiger Zufallsmischungen", *Aufbereitungs Technik*, vol. 17, no. 11, pp. 549–556, 1976.

[SOM 82] SOMMER K., "Wie vergleicht man die Mischfühigkeit von Feststoffmischern?", *Aufbereitungs – Techniks*, vol. 23, pp. 266–269, 1982.

[SPI 92] SPIEGEL M.R., *Formules et tables numériques*, McGraw Hill, 1992.

[STA 54] STANGE K., "Die Mischgüte einer Zufallsmischung als Grundlage zur Beurteilung von Mischversuchen", *Chemie Ingenieur Technik*, vol. 26, no. 6, pp. 331–337, 1954.

[STA 64] STANGE K., "Zur Beurteilung der Güte einer Mischung aus körnigen Stoffen bei bekannten Siebdurchgangslinien der Komponenten", *Chemie Ingenieur Technik*, vol. 36, pp. 296–302, 1964.

[STA 67] STANGE K., "Genanigkeit der Probenahme bei Mischungen körniger Stoffe Einfluβ des Gewichts von Einzelproben", *Chemie Ingenieur Technik*, vol. 39, nos. 9–10, pp. 585–592, 1967.

[TRA 76] TRAHAR W.J., WARREN L.J., "The flotability of very fine particles – a review", *International Journal of Mineral Processing*, vol. 3, pp. 103–131, 1976.

[TSA 94] TSAI W.-R., LIN C.-I., "On the mixing of granular materials in a screw feeder", *Powder Technology*, vol. 80, pp. 119–126, 1994.

[UHL 66] UHL V.W., GRAY J.B., *Mixing, Theory and Practice*, Academic Press, 1966.

[ULB 97] ULBERT Zs., SZEPVOLGYI J., GYENIS J. *et al.*, "Modelling and simulation of particle mixing in gravity and pneumatic mixing tubes containing helicoidal mixer elements", *Récents progres en Génie des Procédés*, vol. 11, no. 51, pp. 299–306, 1997.

[VAL 67] VALENTIN F.H.H., "The mixing of powders and pastes: some basic concepts", *Transactions of the Institution of Chemical Engineers*, vol. 45, pp. CE99–CE104 and 106, 1967.

[VAN 70] VAN HEUVEN J.W., BEEK W.J., "Lois d'équilibre dans la dispersion turbulente liquide–liquide dans les récipients agités", *De Ingenieur*, vol. 82, pp. 51–60, 1970.

[WAN 74] WANG R.H., FAN L.T., "Methods for scaling-up tumbling mixers", *Chemical Engineering*, pp. 88–94, 1974.

[WAN 77] WANG R.H., FAN L.T., "Stochastic modeling of segregation in a motionless mixer", *Chemical Engineering Science*, vol. 32, pp. 695–701, 1977.

[WEI 90] WEINEKÖTTER R., DAVIES R., STEINCHEN J.C., "Determination of the degree of mixing and the degree of dispersion in concentrated solutions", *Proceedings of Second World Congress Particle Technology*, Kyoto, Japan, 19–22 September 1990.

[WEI 95] WEINKÖTTER R., REH L., "Continuous mixing of fine particles", *Particle & Particle Systems Characterization*, vol. 12, pp. 46–53, 1995.

[WIL 63] WILLIAMS J.C., "The segregation of powders and granular material", *Fuel Society Journal*, vol. 14, pp. 29–34, 1963.

[WIL 72a] WILLIAMS J.C., RAHMAN M.A., "Prediction of the performance of continuous mixers for particulate solids using residence time distributions. Part I. Theoretical", *Powder Technology*, vol. 5, pp. 87–92, 1971/1972.

[WIL 72b] WILLIAMS J.C., RAHMAN M.A., "Prediction of the performance of continuous mixers for particulate solids using residence time distribution. Part II. Experimental", *Powder Technology*, vol. 5, pp. 307–316, 1971/1972.

[WIL 97] WILMS H., "Performance of gravity flow blending silos", *European Congress on Chemical Engineering*, Florence, Italy, Part 2, pp. 971–974, 1997.

[WU 71] WU S., "Calculation of interfacial tension in polymer systems", *Journal of Polymer Science Part A: Polymer Chemistry*, vol. 34, pp. 19–30, 1971.

[ZLO 73] ZLOKARNIK M., "Rührleistung in begasten Flussigkeiten", *Chemie-Ingenieur Technik*, no. 10a, pp. 689–692, 1973.

[ZOG 76] ZOGRAFI G., TAM S.S., "Wettability of pharmaceutical solids: estimates of solid surface porosity", *Journal of Pharmaceutical Sciences*, vol. 65, p. 1145, 1976.

Index

Printed in the United States
By Bookmasters